WATT'S PERFECT ENGINE

Watt's Perfect Engine

Steam and the Age of Invention

Ben Marsden

Revolutions in Science
Series editor: Jon Turney

 Columbia University Press   New York

Columbia University Press

Publishers Since 1893

New York   Chichester, West Sussex

Copyright © 2002 Ben Marsden

First published by Icon Books Ltd., Duxford

Library of Congress Cataloging-in-Publication Data

A Complete CIP record is available from the Library of Congress.

ISBN: 978-0-231-13172-8 (cloth)

Columbia University Press books are printed on
permanent and durable acid-free paper.

Printed in the United States of America

c 10 9 8 7 6 5 4

Contents

LIST OF ILLUSTRATIONS

ACKNOWLEDGEMENTS

In writing this book I have drawn upon the excellent work of historians of science and technology, and my first acknowledgement must be to them. Gudrun Richardson and Carol Morgan provided information about Watt's connections, some posthumous, with the Royal Society of London and the Institution of Civil Engineers. Yakup Bektas, John Christie, Graeme Gooday, Christine MacLeod, Dr Eiju Matsumoto, John Reid, Bill Scott and Claire Taylor offered valuable assistance at various stages. Simon Flynn and Jon Turney have been patient, constructive and good-humoured editors, aware, like Roebuck, that it takes time to make things perfect, or even imperfect. Friends, family and colleagues have supported in ways they cannot fully comprehend. Amalia, Rainer and Pat (who read the final draft) get particular and heart-felt thanks. Watt was right about music, which I gave up to finish the book, but thanks, anyway to Fabienne, the Special Trio and Michael (the key-holder) for keeping me saner at a tricky time. I dedicate this book to my loving nieces, Daisy, Rosie and Milly.

Ben Marsden
Aberdeen
July 2002

Introduction:
What's Watt?

Ask any schoolgirl who invented the steam engine and she'll probably answer 'Watt'. But few people now know how a steam engine works, or who James Watt was. We might hazily recall a story about a steaming kettle, just as we think of Newton's apple. We might even pick Watt out as the archetype of the Scottish engineer. With the age of steam long gone, living in the age of information overload, we cling to the fact that Watt invented the steam engine. And every schoolboy knows that although Britain missed the cataclysmic political upheavals of America and France, it had a bloodless but no less profound revolution in industry. The throbbing heart of that transformative Industrial Revolution was James Watt's steam engine.

So 'Watt' has become synonymous with 'steam'. But even Watt did not claim to be the steam engine's sole inventor. It would be crazy to suggest that even the most talented and dynamic individual could single-handedly mastermind a steam-engine revolution.

And although the steam engine undoubtedly played a major role in promoting the Industrial Revolution, Watt's 25-year monopoly on his own engines probably stifled innovation at a time of unheard-of economic growth. To make sense of such a paradox, this book presents a picture of James Watt – inventor, merchant, engineer, businessman and philosopher, warts and all. This is a book about the astonishing machines that Watt and his allies developed and disseminated.

*

It takes time to breed an inventor, and Watt was no exception. This Greenock boy needed an intelligent hand and a head full of books to stave off idleness and live a good life. True craftsmanship was hard to come by, and it took the crowded back-streets of Glasgow and the teeming vistas of a great world city, London, to give the teenage Watt the skills he needed to become an instrument-maker. Few succeed in isolation, and it was Watt's friends and family that opened the doors of Glasgow College, where he set up shop, fraternised with natural philosophers, learned to live the life of the experimenter and invested in get-rich-quick schemes.

When Scotland's movers and shakers cried out for industrial reform, Watt wanted improvement. In the thick of the clubs and chatter of the Enlightenment, his project was progress and, like any merchant, his motive was profit. If experiment worked so well for Watt's friend, the Glasgow College chemist Joseph Black, in taming the physical phenomena of heat, maybe the practical art of steam engineering would succumb, likewise, to sustained and systematic experimental analysis. With that full-frontal attack, perhaps Watt could make his fortune with the idea of the steam-powered carriage dangled before him by the ambitious undergraduate John Robison.

This fantasy was a far cry from the massive coal-guzzling Newcomen fire engines laboriously pumping water in mines throughout the industrial world. But even there, there was so much scope for improvement. When Watt set himself to fix a model of Newcomen's design, just one of a series for the amazed edification of College students, he despaired of making it work without waste. Swallowing steam to little effect, it was clearly inferior to its full-scale counterparts. And yet in analysing its action, desperate to economise on steam, Watt conceived of an idea that really threw a spanner in Newcomen's works.

He asked himself if he could make a 'perfect' steam engine. The *ultimatum* in design, it took a minimum of steam to give a maximum of power. Measured against this perfect standard, real engines like Newcomen's had little chance. Only by going back to the drawing board and introducing the 'separate condenser' would Watt's perfect engine even begin to move from fantasy to reality. It was so simple in theory and yet so hard to bring off in practice. Scaling up from play-thing model to proper machine, he needed to stretch and twist the fabric of the Newcomen engine, piling contraptions on contrivances, obsessed with banishing waste.

Re-inventing steam meant re-inventing and re-arming James Watt. Experimental engines were costly, and Watt was not a wealthy man. Joseph Black was a canny investor in new projects, but for the right resources Watt turned to John Roebuck. With Roebuck's patronage, James Watt, quadrant-maker and merchandiser of fiddle strings, taught himself to be a steam engineer and to act big. He negotiated the hurdles of patent law to protect his ideas. It was all going so well until Roebuck's funds evaporated.

As Roebuck lamented his loss, Watt left Glasgow for Birmingham and the big-thinking buckle-maker Matthew Boulton. He had the unbounded enthusiasm and superlative resources to counterbalance Watt's depthless despond and compulsive mechanising. To fashion an engine for industry took cash, capital and credit. It took perseverance, propaganda and politicking. It meant quick thinking, opportunism and cool nerves. With rivals on every street corner, crafty pirates forgetful of the etiquette of invention, and workers keen to drop a mechanical hint for a dram or two, any ideals Watt might have had about human nature quickly vanished in a cloud of impotent exasperation and legal action. Making a perfect engine meant entering an all-too-imperfect world.

It was not all conflict. Watt found convivial company in the Lunar Society of Birmingham, in which captains of industry, potters and poets, chemists and canal enthusiasts fed all manner of personal hobby horses, and, every Monday afternoon, changed the world. Boulton won battles by networking and it was his surgical lobbying that secured Watt's monopoly for a quarter-century, buying time to make a pump into a factory workhorse, to clear debts and, eventually, to make Watt and Boulton more wealthy even than that boy with the kettle might have dreamed.

Selling started in Cornwall, where mine-owners wanted deep and dry mines but low fuel bills. Making a market, so much vaster than the mines, in the mills and factories meant transforming the engine from a glorified pump to a near 'universal' source of power. That meant fixing it to get rotating power, rating it in 'horsepower' to tell the mill-owners what they were getting for their money, and making it self-monitoring and self-regulating, replacing the unruly middle man with a direct, uncomplaining and intelligent mechanism.

Buffeted on all sides, Watt was trouble-shooter, philosopher and spy-master all at once, learning to make the engines useful but stopping a free-for-all when managers began to covet them, entering a world of science itself peppered with controversy, instructing engineers with regimes of maintenance while despatching friends to watch incognito for technological plagiarism. Jaded and worn down, Watt poured cold water on radical steam-engine innovations threatening the company of Boulton & Watt, and summoned little enthusiasm for the unproven projects of men younger than he who wanted to take a risk on a long-term prospect. Now that the steam engine had survived the turbulence of its adolescence, it was better for James Watt to don long trousers as gentleman philosopher, the only garb suitable to the man increasingly championed as saviour of steam.

Watt's new engines would spread from Birmingham to Cornwall, from continental Europe to the plantations of the West Indies. Steam's progeny reformed factories as society and politics reeled; steamships quickened the pulse of commerce; locomotives annihilated space and time; in sum, steam augmented human happiness in an 'age of steam' – or so the commentators bragged. Never mind those steam-carriages, hamstrung on steep roads; or the endless stream of contenders jostling to supersede steam altogether. Where steam stood for progress, might the steam engine itself, in an age of steam, contain the foundations of its own destruction?

And finally, there were the monuments to Watt and the myths of machine and man that, in making an icon, also struggled to resolve two profound paradoxes. The engine of James Watt, that greatest boon to manufacturing might, emerged under the protection of the near monopoly of Boulton & Watt; yet that monopoly emasculated the technological opposition. The man James Watt, and his partner Matthew

Boulton, exercised the qualities of mind and action required in the business of steam, to bring off a transformative revolution in technology; yet some, at least, of those qualities were hard to condone by those looking for a perfect individual to match a perfect machine. The necessary re-imagining of James Watt began early, and once he died 'Wattolatry' was rife. His noble features immortalised in stone in a vast sculpted monument in Westminster Abbey, Watt looked on, unblinking, as an age of cultural poise gave way to the cut and thrust of a steam society, powered for progress. Now, as a universal unit of power, Watt haunts the language of science and glows discreetly in every electrical light.

For at least a century, the steam engine was the epitome of success. But success is a rare and precious achievement. There are dead-ends in this story, just as there are difficulties surmounted. This is a story of a man forewarned of failure. Jonathan Swift's famous satire on the Royal Society of London in *Gulliver's Travels* (1726) had unworldly Fellows stupidly searching for sunbeams in cucumbers. Watt spent the 1760s and 1770s failing to extract alkali from sea-water. And we should remember that when his foul-mouthed friend Hutton congratulated him on an engine with 'an equitable motion & great power', guaranteed to 'raise his fame yonder', he was talking about the now-forgotten 'steam wheel', not the steam engine. Boulton wanted power for his factory and found more than he can have imagined. Boulton's friend Josiah

Wedgwood wanted power for his potteries. All he got was Erasmus Darwin's 'vertical windmill'.

With work, steam worked. But to understand how, and how Watt took steam from the kettle on the hearth to the mines and mills of the industrial heartlands, we need to understand something about the culture that shaped him – and that he came to shape. How did James Watt get from the shipyards of Greenock to the tips of our tongues? To find out, we begin our story in Scotland.

· CHAPTER I ·

BREEDING AN INVENTOR

Head and Hand

Greenock was a small seaport and ship-building town on Clydeside, 25 miles from Glasgow in the west of Scotland. When James Watt was born there on 19 January 1736, he entered a lower-middle-class family that had been active in the mathematical arts for generations. His grandfather, 'professor of the mathematicks' Thomas Watt (1642–1734), had taught navigation and surveying. His father, also James Watt (1698–1782), chopped and changed, as he had to in a small town, dabbling in anything from cabinet-making to ship-building. He made a good living supplying the navigators of the local herring fleet with nautical goods and mathematical instruments. Greenock remembered him as a resourceful merchant, a trusty town treasurer and a chief magistrate to be feared.

Watt was hardly cut out for the rough and tumble of this noisy, smelly and frantic environment. Four older brothers and a sister had died in infancy and

Watt himself was delicate, plagued by terrible headaches that fostered his legendary moodiness. In the early eighteenth century it was common for mothers who were respectable, or were trying to be, to nurture the impressionable intellects and spiritual sensitivities of their offspring. And so Watt's mother, Agnes Muirhead, taught her sickly child at home. Although neighbours found her eager to impress, harking back to her elevated ancestors, she imbued her son with the dry and unyielding truths of the Scottish Presbyterian Church, with its horror of idleness and waste.

Later, Watt studied at the local Greenock school under the watchful eye of its proprietor, Mr Adam. It was not a happy experience. He was a sickly, boring and spiritless school-fellow. Whether, at the age of thirteen, he really was obsessed by the kettle boiling on the hearth, as his imaginative cousin and companion Marion Muirhead so insistently recalled in later years, we shall never know. We do know that Watt started to study geometry and immediately warmed to it. The staple diet for tyro mathematicians was still Euclid's *Elements*, over 1,500 years old, but it was a discipline for wavering young minds, teaching them straight thinking. His ability at maths and his appetite for the natural philosophy of Isaac Newton was no surprise. Practical Newtonian science was popular and fun – despite its terrifying promise to reduce nature, and with it human action, to law-like predictability. And there was a more immediate prize.

Illustration 1. Not quite Watt's cup of tea? According to a popular story told many years later by his cousin Marion, and subsequently embellished by generations of storytellers, the thirteen-year-old James Watt was inspired by the force of steam gushing from a boiling kettle to invent the steam-engine. Here, his time-conscious aunt looks on, unable to appreciate the momentous 'first experiment' of this romanticised infant genius. (Source: 'First steam experiment of James Watt (1736–1819)', engraving, English School, early nineteenth century. Private collection/Bridgeman Art Library.)

Watt's father, like his uncle and grandfather, was immersed in bills of sale, calculations of profit and loss, and surveys of land. His son and heir had every incentive to grapple with his dividers and master his equations.

When Watt later moved up to the grammar school

at Greenock, he learnt a little Latin and a smattering of Greek. Here, too, was the basic linguistic kit that would allow Watt to explore modern languages – not for fine words but for nuggets of usefulness hidden in practical literatures. These again were essential tools for instilling mental discipline, exemplifying the best in thought and deed revealed in classical civilisations, signalling his social class and his aspirations to transcend it. Watt's best friend at the grammar school was Andrew Anderson, brother of one John Anderson (1726–96), future professor of natural philosophy at Glasgow College (founded in 1451). Friendships were crucial in a society in which connections counted.

Away from all this book-learning, Watt found some useful distractions. His father had a ship-building yard on the south bank of the Clyde, and a workshop where James would tinker with the tools, slaving away at his very own workbench and raising a sweat at a small forge erected for his amusement. At school he used his head for maths; at home he read avidly, ploughing through the books left by his grandfather and his uncle John. But once he left the library for the workshop, he showed that he had an intelligent hand. He took to the workshop completely. He was diligent and he persevered, making and re-making all kinds of models, from cranes to capstans. Soon he was good with wood and a master with metal, turning out delicate work too fine for practical use. But these mini-masterpieces were testimony to his craftsmanship, tangible tokens outbidding any paper qualification he could obtain. This synthesis of head and hand was not unique; but it proved vital when Watt began to think on his next vocation.

Making an Instrument-maker

Although Watt seemed destined to take over from his father, in 1753 business was bad and, even worse, his mother died. His father lost a ship in 1754 and in all this turmoil, when everything was up for grabs, James persuaded him to support his decision to become a specialist instrument-maker rather than a merchant. That meant an apprenticeship in Glasgow. From a market town perched on the Clyde with a port twenty miles distant, Scotland's western metropolis was transforming itself into the 'second city' of the British empire. Trade with North America and the West Indies stocked Glasgow with 'Tobacco Lords' and 'Sugar Barons', keen to invest their new-made fortunes to improve the city, open up its river, and develop new industries like tanning, iron-working, printing and glass-making.

Clutching his gimlets and chisels, sporting a leather apron, stocked with eight night caps and a tartan waistcoat, James Watt arrived there in June 1754. He had much to learn. Watt spent an entire year apprenticed to a master who could teach him to make experimental apparatus, basic musical instruments, and surveying tools. There is a common story that Watt was wasting his time with such teachers, fixing fishing tackle. But a variety of skills and the products that went with them were essential for business success in a place like Glasgow, where single-mindedness in this trade signalled commercial death. Watt often lodged with his mother's brother, John Muirhead. Family connections often opened doors. John Muirhead's son George was professor of humanity (Latin language and literature) at Glasgow College. Through him, James met the professor of natural philosophy, Dr Robert Dick.

Dick was convinced that to get decent training as an instrument-maker, Watt should move on to London, then Europe's leading centre for the trade. In June 1755, the nineteen-year-old embarked on the twelve-day journey by land to the English capital. It was a formidable adventure. Dick gave Watt an introduction to the Edinburgh-born maker James Short, a Fellow of the Royal Society of London. Founded in 1660, it was still the most important scientific society in Britain, even if its prestige had dwindled since Isaac Newton's death, and publication now took second place to clubbability as a criterion for membership. With his business in London, Short had a national reputation as an instrument-maker specialising in reflecting telescopes. But away from the familiar protected environment of Greenock and Glasgow, Watt ran into difficulties. It was presumptuous of Dick to think that Short might take Watt on, but he could still offer help.

Watt needed a master who could, for a price,

inculcate the expert skills – but also the *variety* of skills – that he wanted. London makers tended to be too specialised. Short passed him on to John Morgan, a 'philosophical instrument maker' who produced instruments for science and the practical arts, and one of only five or six makers with the mix of knowledge that Watt demanded. Morgan had premises at Finch Lane, Cornhill, and was best with brass. In July, a little on the sly, he took Watt on as a 'premium apprentice' for a year, speeding his transition from indulged amateur to canny businessman.

Watt used his year with Morgan to learn how to make the instruments that his father's Greenock clientele wanted: simple rules and scales, Hadley's quadrants (for navigators) and portable barometers. He could make them rough and ready for quick and lucrative sale. But the holy grail was precision work-manship, and eventually he too could fashion and fabricate with a degree of accuracy not needed in everyday practice. This was the royal road to quality, and the mark of the specialist maker's prestige. Accuracy delivered reliability, security and longevity. And these promoted the savings so vital to Watt in the coming years.

By Christmas 1755, Watt was already dreaming of his own business. The following spring he was cautious but capable, already on a par with the trained craftsmen, known as 'journeymen', employed by others before they branched out independently. In June 1756, as his year with Morgan came to an end, he prepared a shopping list: £23 12s 6d (perhaps six months' salary for a skilled artisan) to be paid out for hammers and files, tortoise shell and tubes, wire and mother of pearl, and all the other equipment needed to set up as a West-of-Scotland instrument-maker. But soon Watt was suffering (again) from a racking cough and a gnawing pain – whether from the noxious atmosphere, from intense overwork, or simply revulsion at the moral turpitude of London, isn't clear. In July he trekked northwards, ready to start work for real.

The Business of Natural Philosophy

By August 1756 he was home. He stayed there until October, recuperating, helping his father out, and pondering his next move. Watt was finely poised between the certainties of Greenock and the commercial and intellectual prospects of Glasgow. When he visited Dr Dick in October 1756 he walked straight into paying work. A wealthy merchant, Alexander Macfarlane, had amassed a glorious collection of astronomical instruments, which he housed in a special observatory in Kingston, Jamaica. In October 1755 he bequeathed them to Glasgow College, his *alma mater*, but as they made their way to Scotland by sea, they salted up. Dick wanted Watt to unpack them, clean them and put them into good order. He even arranged a room in the College where he could work. With jobs like this, Watt found favour with Glasgow's professors, but Greenock was still a draw. He wavered, spending the spring of 1757 in Greenock failing to make much impact as an instrument-maker, before returning, once and for all, to Glasgow.

Now was the time to capitalise on his schoolboy friendship with Andrew Anderson. In May 1757, Dick died and Andrew's brother John Anderson vaulted from the chair of oriental languages to the chair of natural philosophy to fill the dead man's shoes. Watt's training in Glasgow and London, and his family connections, did him no harm when Anderson and other professors, anxious to patronise a specialised trade with little Scottish support, lobbied to make Watt 'Mathematical Instrument Maker to the University'. Their success certainly enhanced Watt's local profile and gave a kick-start to his business. The arrangement was quite informal, but it was not unheard of. As with the Macfarlane bequest, employment around the College was organised piecemeal. Professors with practical classes often paid artisans to demonstrate or to fix apparatus, knowing that more attractive lectures snared students – all paying fees to them directly. In fact, both of the men most likely to have been Watt's master in Glasgow had been 'laboratory stewards' for the natural philosophy professors.

There is a myth, created inadvertently by Joseph Black (1728–99), that by giving him this appointment the College protected him from the vindictive persecution of petty locals. When he tried to start work in Glasgow, so the story goes, the Incorporation of Hammermen (skilled metal-workers) invoked traditional guild restrictions and prevented him from doing so: he'd been trained in the wrong place (not Glasgow) and for too little time (two years at most). But both of the top candidates for Watt's master in Glasgow were members of the Incorporation. Watt must have known the score: as a 'stranger' seeking to practise, he simply had to pay a small fee for a 'Burgess ticket', if asked. Alas, there was no such adversity, here at least, over which our youthful hero triumphed. In fact, he just went about his business unmolested.

At the end of July 1757, the College allowed him to transfer his tools (among them his 'great lathe') to a small workshop within its precincts. Thanks to Anderson, Watt had College apartments. Since Watt was there for the science teachers, ready to respond to *ad hoc* commissions, to revive Macfarlane's instruments, and probably to help with a newly-founded 'Macfarlane observatory', a College address made sense. It was still a privilege, given the hysterical disputes about space which erupted sporadically in that close-knit community. Here the College bowed to Francis Bacon, for whom knowledge was power, by forging bonds between intellect and industry: Anderson inaugurated so-called 'anti-toga lectures' in the evenings for local mechanics and artisans; and, just as important, the College dedicated space to printers, type-founders – and an instrument-maker. Watt was not the only one dirtying his hands in the halls of learning.

Working in the College gave him an easy *entrée* into the bustling clubs and academies of the Scottish Enlightenment. This was a time when history, philosophy, economics and the human sciences flourished north of the border. During this time, Watt met up with close friends in an 'irregular club', sparring over religion, and mulling over the latest literature. Academic connections coaxed Watt into the company of many a penetrating intellect, jousting and jostling for position in a veritable hotbed of genius. Glasgow College itself, although a tiny outfit, boasted incredible talent.

There was Adam Smith (1723–90), professor of moral philosophy and a man whose reconstruction of economics, anatomising skills in a 'division of labour', promised to revolutionise industrial production. When Watt first met John Robison (1739–1805) in 1756, the man was a talented undergraduate, irascible and aiming high. Although Robison completed his degree and left Glasgow in 1759, their discussions recommenced when he returned four years later, and they continued when he moved on to Edinburgh as professor of natural philosophy in 1774. Their friendship withstood Robison's post-revolutionary rants against French liberalism and his paranoid denunciations of freemasonry (despite Watt's own leanings towards the lodge).

Just as important, when Watt arrived Glasgow had just become the institutional home of Joseph Black. In 1756, the departure of professor William Cullen to Edinburgh, and a typically Scottish re-shuffle of professorial chairs in Glasgow, saw the illustrious bachelor Black appointed to a combined chair of anatomy and botany. Within a year he had moved up to the chair of medicine. A man noted for coolness, he was nevertheless passionate about chemistry, and lectured in it despite there being no independent chair. Compounding medicine and chemistry made excellent sense in Scotland, where cohorts of trainee medics paid up for compulsory chemical science.

While completing his Edinburgh medical degree, Black had already become a skilled 'pneumatic chemist', manipulating materials and apparatus to isolate a new 'kind of air' from common atmospheric air and giving it the name 'fixed air' (later renamed carbon dioxide). Cautious Black may have been, but he courted philosophical controversy in Glasgow by investigating the phenomena of heat, a topic traditionally belonging to the chemist. Through a long series of experiments between 1756 and 1766, Black developed two crucial concepts. The first was 'heat capacity'. Black found that different substances had different 'capacities' for heat: that is, to achieve the same rise in temperature, different substances required different transfers of heat (more, say, for iron than for wood). The second was 'latent heat'. As far back as 1755, Black had been interested in the transfer of heat to or from a body during liquefaction, solidification, evaporation and condensation (changes of state) without any change of temperature showing up on a thermometer. Since heat was clearly transferred but was not revealed, Black thought of this as hidden, or latent, heat. Although reluctant to put these ideas into print, Black presented them to his Glasgow students – an inclusive mix, typical of a university which, unlike Oxford and Cambridge, extolled the virtues of a 'democratic intellect' and welcomed in its crofters and its artisans.

Watt's relationship with the professors was good, and with Black, especially, it went far beyond the traditional one of employer and employee. Black quickly recognised Watt's 'uncommon talents for mechanical knowledge and practice' and his 'originality, readiness and copiousness of invention'. They were soon intimate friends, swapping diagnoses of personal ailments and prophesying plausible technological futures. After all, Black had close ties with the Board of Trustees for Manufactures in Scotland, a body designed to stimulate the national economy out of the doldrums through innovative practical arts.

So Black was no dry theorist. He lobbied hard for laboratory space and equipment. Right from the start, he employed Watt frequently to create and repair apparatus. For Black's science classes there were doors for a new kind of furnace, a 'condensing syringe', and even a 'steam digester' (like a pressure cooker). John Anderson was the main source of Watt's College business, asking him to fix or fabricate everything from pulleys to pumps and air guns to comets (or rather, models of comets). Under the professors' skilful show-manship, this paraphernalia revealed the wonders of heaven and earth brought to order in Newtonian natural philosophy.

Watt divided his time between jobs like this, making or repairing instruments for general sale, and general merchandising. Not surprisingly, his university duties neither occupied all his time nor gave him a steady income. Work done piecemeal for the professors was intriguing and it carried a certain kudos. But each one-off job cost time in honing special skills. Cleaning clocks was no way to make his fortune. It was more bother than it was worth.

It was better to expand the business and to broaden his range, just as he had intended when he first trained in Glasgow. In 1758, after a year working in the College, Watt began to employ assistants to cope with demand. For Black he made an alarm clock to use at home. But the most likely path to profit, safer and more lucrative than College work, was to churn out scales, balances, saleable Hadley's quadrants, and compasses, at first for his father's navigators but later for his own growing clientele (which included Glasgow student David Lyle, who took three of Watt's compasses). By 1764, Watt was employing sixteen workmen. Under the guidance of this driven man, more 'workshop manager' than solitary craftsman, they made more and they made cheaply, 'mass producing' standardised parts, utilising simple tools for cutting, printing and engraving that saved labour and brain-power. Always the aim was to cut out waste – and maximise profit.

Watt had trained as a philosophical instrument-maker, but he soon expanded the business to include musical instruments. He was no fan of music *per se*. His grandfather Thomas Watt had zealously clamped down on musical fun. James decided that music-making was the root of idleness and, years later, made one prospective employee take an oath that he would give it up entirely. Reformed, this man, John Southern (1758–1815), would become one of Watt's most creative and valued assistants. But provincial mathematical practitioners had to be versatile: jacks of many trades and masters of some. Black wanted an organ, Robison wanted a bridge for his fiddle, and the practical astronomy professor Alexander Wilson wanted guitar strings. Who was Watt to argue when profit was in the offing? Many of the tools and materials needed to make and repair musical instruments were to hand; when Watt did not have the skills, he hired in specialists who did. The shop was soon retailing, repairing, or producing violins and bagpipes for Glaswegian amateurs, shipping instruments to London, Edinburgh and Dublin (and sending out parcels far and wide, from Bristol to the West Indies and from Greenock to Quebec). There were cheap and cheerful boxwood flutes, each worth a few shillings – but fetching a little more, perhaps, if they were embellished with élite Parisian designer labels. And strange to say, the stamp of one such maker, Thomas Lot, did surface years later among Watt's tools. Streamlining production was not the only way to maximise income.

All this entrepreneurial activity is a far cry from the popular picture of Watt, a simple isolated craftsman, diligently working away in the leafy groves of academe, oblivious to the wider world. To carry out these grand plans for expansion, Watt needed capital. After a lifetime of hard work, his father had made – and lost – a considerable fortune, so there was little chance there. From 1757, Watt started to borrow money from the merchant John Craig, and by October 1759 they were in partnership. Craig annually lent Watt sums of £100 or more, to be repaid with interest. Only with this injection of funds could Watt expand and diversify the business, take on more journeymen, and open another shop. This one was outside the College, first in Glasgow's Saltmarket and then in the Trongate, and he sold everything there from snuff-boxes to shoe-buckles and from pistols to nut-crackers.

Established in business, it was time for Watt to marry. His cousin Margaret Miller became his wife in 1764. Sometimes she'd help out in the shop. Their son James Watt Jr (1769–1848) was soon being nurtured as Watt's business heir. Now, in 1764, Watt had a choice. He could continue as he was, developing his instrument business, building up capital, and invest-ing in new commercial ventures like Glasgow's Delftfield company, a new industry founded on the

Tobacco Lords' capital to imitate the famous Delft pottery. James Watt, 'Merchant', might have gone the way of his father, ending his life as a pillar of the community, administering justice and dribbling money back into the town in charitable works.

But, as we know, that's not what happened.

· Chapter II ·

Re-discovering Steam

Hero's Toys and the Miner's Friend

Watt had always been interested in mathematics, natural philosophy and chemistry. Once he moved into the College, Black and Robison brought home to this craftsman-entrepreneur the seductions, the rewards, and also the profits of science working hand-in-hand with the practical arts or technologies. Black especially was a walking repository of scientific advice and practical wisdom. But it was Robison, Watt claimed, who encouraged him in 1759 to think for the first time about what might be done with the steam engine. Watt was woefully ignorant about it – but then why should a young instrument-maker be otherwise? He was attracted, though, by Robison's quirky fantasy: perhaps a clever mechanic could harness the power of steam to turn the wheels of a carriage.

That was the start of it all. Of course, Watt did not need to 'invent' the steam engine from scratch. A thousand years ago, Hero of Alexandria was fabricating steam-driven toys. More than a century before Robison tried to excite Watt with stories of steam-spun carriages, men like Solomon de Caus (in 1616) and the Marquis of Worcester (in 1659) had been scheming to make 'fire engines' that could raise water by steam. And just 50 years before Watt set up shop, mechanics racked their brains as they tried to transform fire engines from philosophical toys into machines of profit, when all around they heard talk of a symbiosis between high science and craft technique. One of these technological projectors, the Huguenot Denis Papin (1647–*c*. 1712), found his niche at the fledgling Royal Society of London. Engaged to perform experiments for the Society's gentlemanly fellowship of Christian 'virtuosi', Papin came up with a simple engine in 1690 – and that 'steam digester' which later interested Black.

Others looked for ways to use steam to pump water, either to open up deep mines for profit, or to keep waterwheels turning. There were two main ways of doing this. In the first, the trick is to use steam to create a near vacuum (or empty space) inside a 'cylinder' (a cylindrical vessel). Just fill the cylinder with steam to push out the air, and then sprinkle in cold water. The steam condenses, leaving a partial vacuum. With the right configuration of pipes and valves, the push of the atmosphere (atmospheric pressure) can then be used to push water upwards into the (nearly) empty cylinder. Expel the water through a valve and you have a simple pump. The second, alternative, technique is to use the pressure of steam, over that of the atmosphere, to push against the surface of water held in a reservoir. Part of that reservoir of water can then be expelled upwards through a pipe, and the higher the steam pressure, the higher it will go. In two different ways, sucking or pushing, steam helps to raise, or pump, water. In the first, the atmosphere does the pushing against a vacuum. In the second, the steam pushes against the atmosphere to do the work.

Few of these schemes had much impact until the arrival of Thomas Savery (*c.* 1650–1715). He advertised his 'engine for raising water by fire' in the brochure that gave it its nickname: *The Miner's Friend* (1702). The engine used a vacuum, generated using steam, to 'suck'; and it also 'pushed', using the pressure of steam over and above that of the atmosphere. It was a very temperamental friend, but that was no problem for Savery. By 'patenting' the engine in 1698 he secured a fourteen-year monopoly on 'raising water by the impellent force of fire', and in 1699 that monopoly was extended to 1733. Patents were supposed to encourage investment, to reward the patentee, and benefit national industry. But Savery used his to coerce engineers trying any method of 'raising water by fire' into collaborating or paying a heavy royalty. The Dartmouth ironmonger Thomas Newcomen (1663–1729) was just one cautious inventor who had to work with Savery and then with the speculators who snapped up his patent after he died.

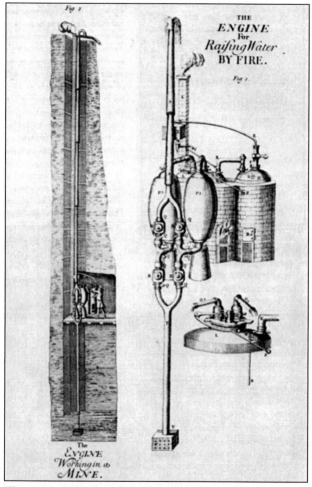

Illustration 2. Thomas Savery's engine for raising water by fire. Although the engine was troublesome in practice, the patent for it was the basis of a lucrative 30-year monopoly on pumping by steam. (Source: Thomas Savery, *The Miner's Friend* (1702).)

As a complete outsider in the fire-engine business, all this was new to Watt. Robison would scan the libraries to satiate Watt's hunger for knowledge of steam. He found modern classics like John Desaguliers' *Course of experimental philosophy* (1734–44), a book encapsulating the author's famous lectures on the practical arts. Inside its covers, theories of heat mingled with minute descriptions of engines ancient and modern. Watt also consulted the technical author Forest de Belidor's *Architecture hydraulique* (1739–53), a weighty tome spiced with accurate descriptions and stories of machines too distant – or fanciful – to observe in action. The literature of power engineering was stuffed with clues about exotic and enticing mechanical futures. Maybe, just maybe, human ingenuity could tweak the known and the mundane to conjure up power for free in a perpetual motion machine. An imaginary 'book of blots' cautioned unsuspecting mechanical tyros to avoid such pipe dreams and walk on by the blind alleys that led only to dismal failure and moral degeneration. All of this urgent mechanical babble pointed beyond the horizon to a place where the stupendous natural power of steam just might be harnessed for the good of humanity – and the good of James Watt.

Reading alone could not provide all the answers. Even the best books by philosophers on practical matters could be unreliable. Watt came to the conclusion that even a good Newtonian like Desaguliers, who prided himself on learning through experiment, had his facts about steam wrong. This encouraged Watt to copy Black and try his own homespun experiments to penetrate nature's mysteries – here, the mysteries of steam. Watt was well placed in his workshop, with materials to hand, to do this – torturing nature in search of steamy novelties. Simultaneously he looked to practical applications, early on mocking up a little model engine with tin-plate cylinders designed, he hoped, to turn the wheels of a carriage and then, maybe, to revolutionise transport, and so on. As (almost) ever, practical difficulties emerged. Neither Robison nor Watt could find solutions, there were other more pressing concerns – so they consigned that scheme to the rubbish bin.

With Robison dispatched to Canada, Watt continued his steam experiments on and off until 1763. Using the balances, barometers and 'mercurial manometers' that he made himself, and thermometers that he bought in, he charted the complex relationship between steam pressure and temperature. In 1761 or 1762 he 'tried some experiments on the force of steam in a Papin's digester' (like the one he provided for Black). Remarkably, he found that steam at very high pressure from the digester entering the syringe and pushing against a rudimentary piston could lift a weight of no less than 15 pounds. Scribbled diagrams in Watt's 'Notebook of experiments on heat' (dating from between 1764 and 1766) show that he really *did* use that kitchen kettle (what else?) when he needed steam. In one experiment he 'took a bent glass tube & inverted it into the nose of a tea kettle', putting the other end in cold water so that he could gauge how much heat steam really contained.

Watt later claimed that in order to make a proper steam engine, beyond all this experimenting and model-making, he needed to understand the *principles* of steam. His increasing business activities left little

Illustration 3. Watt's kettle. Watt frequently used a kettle in his experiments, and here he made a thumbnail sketch of the simple apparatus with which he measured the quantity of steam needed to heat cold water to boiling point. Joseph Black explained Watt's result in terms of his concept of 'latent heat'. (Source: Watt's 'Notebook of experiments on heat' (1764–6), courtesy of Birmingham City Archives.)

time for such unprofitable investigations, however, especially when there was barely more than a glimmer of hope that something really worthwhile would come of all this tinkering. Something more concrete would focus Watt's attention.

Learning About the Newcomen Engine

Late in life, anxious to remind the world that he was no mere tinkerer but a *bona fide* man of science, Watt claimed that it was only in 1763, after intense study, that he finally clarified the *principles* he took to be the basis of his steam-engine improvements. The records show a more complex story: Watt scouring the literature for useful nuggets hinting at steamy futures; Watt beavering in his workshop, mocking up demonstration models and isolating steam's properties to understand them better; Watt happy at Robison's return, also in 1763, and resuming their collaboration. And of paramount importance in Watt's quest to conquer steam, and to build a working engine as 'perfect' as he could make it, was no abstract idea but a battered and broken model of an engine that was anything but perfect.

Along with its buildings, furniture and books, Glasgow College coveted its collections of models, 'demonstration' apparatus, and specimens. Scottish professors and their assistants relied heavily upon this 'material culture', exhuming mechanical exhibits from cluttered store-rooms and wheeling them out into classrooms where they taught useful mathematics, natural philosophy, and chemistry – especially in their practical or 'experimental' guises. Since natural philosophy was a core component of the foundational arts curriculum, their lectures were well attended by audiences looking to be entertained and morally uplifted rather than professionally prepared. Skilled lecturers re-animated these inert devices and contraptions, mobilising them to celebrate the fun, the spectacle, and the utility of science.

College professors in Edinburgh, Glasgow and Aberdeen bought up sets of apparatus or had their own designs put together by local craftsmen – hoarding the end products as material and intellectual capital. With a well-thought-out collection of demonstration apparatus, and skills of exposition practised over the years, a lively professor could entice students from all levels of society to part with money and to attend his classes. Black's lecture demonstrations – and thus earning power – were reputed to be extraordinary. Issues of ownership usually came to a head when an old professor finally died, or an active one shuffled off to a new chair elsewhere, and there was a scramble for a new appointment.

Among the models hoarded in Glasgow College, ready for demonstrations in Anderson's natural philosophy classroom, there was one particular fire engine. It was a simplified, and miniaturised, scale model of the engine that Thomas Newcomen had spent fifteen years developing. He had made his first

Illustration 4. Thomas Newcomen's 'ENGINE for Raising Water (with a power made) by FIRE' (of 1717). Unlike Savery's machine, this vast installation had a real commercial impact and was common in collieries and mines from the 1730s. Watt could see an almost identical image of the Newcomen engine at Griff, Warwickshire, in the *Course of experimental philosophy* (1734–44) by John Desaguliers. (Source: engraving by Henry Beighton (1717).)

full-scale engine in 1712 for the Earl of Dudley's colliery, in sight of Dudley Castle. Since many prominent land-owners adopted the engines to revive flooded pits or open up new ones, especially after Savery's patent expired, Belidor described their innards in detail. Desaguliers had been responsible for setting up Newcomen engines, particularly around London, and very likely for the upmarket model in the apparatus collection of George III, so he certainly wrote about them with authority. Although it was not easy for Watt to examine one directly – Glasgow had no full-sized Newcomen engine until 1764 – he turned to these books and to the College's model.

These Newcomen engines were in fact simple to describe. Imagine a vertical hollow cast-iron cylinder with an open top. And inside it is a piston (a circular disk at the end of a rod) moving up and down. This piston is connected by a series of chains to a quadrant (a quarter of a circle). The quadrant, in turn, is fixed to the end of a heavy wooden beam. The beam is then pivoted half way along, like a see-saw. The other end of the beam is connected to a pump. The parts of the engine are balanced so that the piston in its natural position of rest is at the top of the cylinder. In order to start the engine, air is pushed out of the cylinder by letting low-pressure steam from a boiler enter from the bottom. The supply of steam is then cut off. Water is 'injected' (or sprayed) into the cylinder and the steam quickly condenses, leaving a partial vacuum below the piston. The pressure of the atmosphere will now push downwards against the top of the piston and with only a vacuum beneath it, the piston falls. (This pushing action of the atmosphere makes the Newcomen an 'atmospheric engine', strictly speaking, rather than a 'steam engine': the steam is essential in

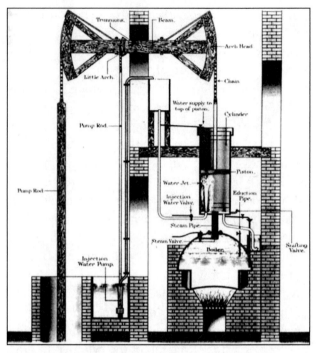

Illustration 5. The workings of Newcomen's atmospheric engine. Although simple in theory, in practice the engine required a complex, integrated array of apparatus: a rocking beam and chains to transmit motion from the piston to the pump, a boiler to generate steam, a tank of water for 'injection' into the air-tight cylinder, and the 'snifting valve' to get rid of unwanted air and water. (Source: J. Farey, *A treatise on the steam engine* (1827).)

creating the vacuum but it never actually pushes the piston to generate power.) Now, as the cylinder end of the beam moves down, the other end rises and so works the pump. Once that's done, the weighting of the machinery takes the piston up to the top again and the cycle of actions, called a 'stroke', repeats.

Although this explanation makes the machine seem easily comprehensible – more so than many modern technologies with their inner workings concealed within black boxes – in practice, much more was needed to make a large-scale engine which functioned reliably whatever the weather. There had to be a water supply, a cistern (or tank) for the injection water, a small pump to get the water up into the cistern, and pipes to spray it into the cylinder. There was a valve to regulate this spray of water, and another, called the 'snifting valve', to get rid of the surplus air that clogged up the cylinder. (It sounded like a man with a cold.) To make the cylinder air-tight there was leather packing inside and a pool of water over the top of the piston ('water packing'). There was a boiler, looking like a brewer's copper, and that had to have a safety valve, even though steam was rarely ever used at pressures more than a few pounds per square inch above that of the atmosphere (approximately fourteen pounds per square inch). The beam had its own (hopefully) robust gearing. The first engine was 50 feet high and required a heavy masonry 'house' to protect, support and steady it. Newcomen had done his best to automate the action of the engine but there were still people, too, in attendance. Engineers and stokers were forever on call to feed these temperamental mechanical beasts or to coax them into life.

These Newcomen engines could also be tricky to construct and control. Initially, getting them up and running was no easy task. Engines made by skilled craftsmen in England needed trained experts to 'finish' them on site, not least because every mine had its idiosyncrasies. This was especially true when engines were 'erected' in new countries, like France, from the 1720s. Even experienced engineers, like Desaguliers' acolyte Martin Triewald, struggled hopelessly with skill shortages, poor materials and even bad weather when he put this technology on trial in Sweden in 1726. It was a long time before this could truly be called a simple solution to the drainage problems of deep mines.

Ultimately, the Newcomen engine became a familiar part of the industrial landscape, changing little in basic design between 1720 and 1760. It was not by any means a universal source of power. A very special assemblage of machine and man, it pumped water in coal mines where fuel was in ample supply, close to hand, and cheap. Thanks to these engines, mines could follow rich seams ever deeper, balancing danger against profit. Newcomen's machines were also installed in the deep tin mines of south-west England, and especially Cornwall. Without pumps working round the clock, these mines were simply not viable. But there was one crucial difference. In Cornwall, pumping was expensive since fuel was scarce. Although owners welcomed the Newcomen engine for its double-edged promotion of safety and deep working, they were constantly on the lookout for improvements – or even radical alternatives.

Enter, then, James Watt. Robison had suggested he look at steam, and there was the pipe-dream of the steam-carriage. But this was a more specific problem – potentially more rewarding, but likely to be equally intractable. What could be done to 'improve' or supersede a reliable technology which, although tried and tested for more than 40 years, was increasingly seen as anything but 'perfect'? The model Newcomen engine was a valued part of the College's collection at a time when its full-scale counterparts were by far the dominant fire engines employed in industry. This was the best concrete example of steam in practice. If Watt was to do better, he must first understand the workings of the Newcomen engine – and then supplant it. He can have had little idea that to do that would soak up his energies for the next twenty years.

· Chapter III ·

Re-inventing Steam

Re-modelling Newcomen: Watt's Perfect Engine

The best account of what happened next has been provided by the historian Richard Hills, and here I follow his analysis. There was a problem with the model Newcomen engine. Although interesting to the class, as a fine example of mechanics at work in the mines and promoting industrial expansion, it did not work well. A sojourn in London to be looked over, or perhaps even reconstructed, by the instrument-maker Jonathan Sisson (*c.* 1690–1760) did not improve its health. In 1760, the College's governing Senate offered Professor Anderson a hefty grant of £2 to get it back.

Watt was given the task of making it do better. There is a popular myth that the engine was quite seriously broken and that, as Watt himself said, he needed to repair it. Certainly, Watt was often taking on commissions to repair apparatus for Anderson: in 1762 he received 6d for 'Mending a Barrometer', for

44

Illustration 6. The model Newcomen engine which Watt repaired and modified in the winter of 1763–4 for the Glasgow College natural philosophy class of John Anderson. Watt wanted to reduce the wasteful consumption of steam from the boiler (bottom left) in the cylinder (top left) when, at each stroke, cold water from the cistern (top middle) was injected to create a vacuum. Experiments with this and other models soon led him to re-evaluate the design of full-sized engines. (Source: Hunterian Museum, Glasgow University.)

example. But in this case by far the more interesting part of the job was to *improve* the action of the machine. In the winter of 1763–4, Anderson turned to James Watt and added to his list of tasks that of making the engine work *well*.

By the end of 1764 (and probably months before) Watt was experimenting with the Newcomen model, with its tiny brass cylinder of two inches by six and a simplified set of valves. No solitary worker, Watt called on men he trusted, like James Fleming, to repair or provide parts such as the cistern with a tin lid that held the condensing water. There was more to it than simply polishing brass or replacing existing parts. Once it was up and running, Watt found that although the nine-inch 'boyler' which Fleming probably fixed for him was 'apparently quite large enough' relative to the engine's diminutive scale, it really could not supply enough steam to keep it going long enough for a convincing demonstration to eager young students. Watt conceived of the problem not as a question of fuel used, or even of power produced, but of steam consumption. Quite literally, the engine kept running out of steam.

We should remember that, at least at first, rather than thinking to improve the full-scale Newcomen engine, Watt was concerned to see why a model did not work as well as the 'real thing'. He tried to make more steam, boiling the water more violently, but that just made the engine grind to a halt. Then he put the cistern higher up so that water would be injected at higher pressure, but he found that he needed a huge quantity to get the required vacuum. Since it was not his engine, he was wary of making radical changes, so he returned it to the College (by 1766 at the latest) and pocketed £6 for services rendered.

Robison observed how that original model had been like a 'fine play thing' to Watt, but following the experiments of the winter of 1763–4 and afterwards, it became an object of science. Watt claimed that although he began this work as 'a mere mechanician' and an amateur in science, it transformed him into a natural philosopher, bringing him closer to Black. Increasingly he aped the man of science, recording results in a special notebook, brainstorming ideas, and finding practical ways to study phenomena in isolation. At the same time he dissected and re-created the engine, allowing experimental phenomena and practical discoveries to spark off each other and cross-fertilise.

For the most significant modifications of the

Newcomen engine, he worked with his own models. Watt saw that there were many ways in which a small model engine differed from a full-scale one. He could have learnt from the speculations in Isaac Newton's *Opticks* that small bodies tend to lose heat more rapidly than large ones of the same shape and material. A smaller cylinder would lose heat more quickly than a larger one, because proportionately it had a larger cooling surface. Nothing much could be done about that. Also, Watt considered that the brass cylinder of the College's model might 'conduct' heat (receive it, or give it up) more quickly than the cast-iron cylinder of a full-scale engine, which soon became caked with 'a stony crust' that did not transmit heat. Why not vary the materials from which the engine was made? Wood, for example, was cheap, and in Scotland it was plentiful. He made a model with a wooden cylinder about six inches by twelve, but the steam tended to warp it. Material reality intervened. It looked like metal cylinders would have to do. Furthermore, he found that the amount of steam required to heat up and completely fill the wooden-cylinder model was much greater than it should have been according to Desaguliers' figures for large-scale engines.

Watt's major concerns were these: to establish how much steam the engine consumed in each stroke; to find a way of reducing that consumption to a minimum; and to ensure that the vacuum produced by condensation was as good as possible. But what was that minimum? The very least that might be needed was one cylinder full of steam for each stroke. And how good could that vacuum be? Ideally, it would be a total vacuum, with no residual steam, air or water left in the cylinder.

An engine using only one cylinder of steam and producing a total vacuum would be 'perfect'. Such a 'perfect steam engine' (as Watt called it) represented the ultimate goal of the steam engineer. It was a standard by which real engines could be measured, or against which the progress of prototypes could be charted. In September 1765, Watt told his friend James Lind that he had already 'tried' a 'small model of my perfect engine'. With it, he expected 'almost totally to prevent waste of steam, and consequently to bring the machine to its *ultimatum*'.

Watt was not the only one to think of such perfect standards. Navigators wanted a perfect time-keeper to enable them to determine longitude and, by a neat coincidence, Robison had just returned from a voyage to Jamaica designed to test the best that was on offer: John Harrison's famous chronometer 'H4'. In a penetrating analysis of the comparative powers of wind and water mills, the élite civil engineer John Smeaton (1724–92) talked about a perfect water mill, by which he meant one that could pump enough water to keep itself in constant supply. Real mills had no chance of achieving this impossible goal, but it set a standard against which they could be evaluated. French engineers estimated that 'undershot' wheels were harnessing only 4/27 of the available work of the stream, and as recently as 1760 Smeaton had calculated that, with overshot designs (where the water first hits the wheel at the highest point), this figure could be vastly improved to two thirds. Watt's 'perfect engine', likewise, set a target for improvement.

Now he had to calculate how many cylinders of steam his engine actually did consume for each stroke, knowing only that it was at least one. To find this out he had to measure two things: the amount of boiling water needed to make a volume of steam equal to that of the cylinder; and the amount of water evaporated in the boiler over a certain time as the engine was running. The second figure was easy to come by: he simply had to stoke up the boiler, weigh the water inside it, leave it to run for a set time, and then weigh what was left. Then it was easy to calculate the amount of water used up.

The first was more difficult. Watt doubted Desaguliers' published estimate that, in making steam at atmospheric pressure, boiling water was expanded 14,000 times. At this point he turned to Black. Since Watt, Robison and Black all gave slightly different accounts of what happened next, we will never know just how great was Watt's debt to the College professor. In one experiment, Watt followed Black's instructions, and using somewhat crude apparatus he boiled water in a flask, took careful measurements of weight (typical of a chemist), worked out how much water was evaporated, and concluded that water expanded to roughly 1,600 times its original volume as it turned into steam.

When the engine was running it consumed all the steam produced by the boiler, so, from these two figures, Watt could work out how much steam was consumed in each stroke. The answer was not one but several cylinders of steam – three, according to one estimate. Watt's engine was anything but perfect. In comparison to the 'perfect steam engine' it was as wasteful as those inefficient waterwheels, except where they wasted water, Watt's engine wasted steam – and coal. Since a fixed amount of coal boiled up a fixed amount of steam, an engine consuming three times the steam burnt three times the coal – and that cost money. To approach perfection, Harrison struggled with his chronometers and Smeaton optimised his mill design. How could Watt approach his *ultimatum*?

He could do it by locating – and removing – all possible causes of waste. The waste of steam and fuel grated. Watt's religious and social environment was one in which all forms of waste were abhorrences to be rooted out by self-discipline and intellectual application. This attitude permeated Watt's doings, from merchandising to mechanics. Between 1757 and 1763 he kept an 'Inventary [*sic*] of the Goods Money Debts belonging to me James Watt junr also what I owe others'. This 'Waste Book' minutely recorded income (an 'old Big Coat' sold for eight shillings) and expenditure ('1/2 lb. Quicksilver – 2.9') in the hope of future savings. Even a trip to the dentist had its price ('Physick & drawing a tooth': 3/6) – but could he do better elsewhere? Asking his father to send supplies, he insisted: 'not by Willie Crawford as he's an extravagant rascal'. In a life monitoring income and expenditure, and shunning excess, Watt wanted an engine with clear inputs and outputs, free from avoidable waste.

It is sometimes suggested that Watt was motivated from the beginning to make a more powerful engine, or that he was shocked by the model's waste of 'energy'. His syringe experiments indicated that steam, especially hot steam at high pressure, could be harnessed mechanically to produce a great 'effect' (to use Watt's term) or 'mechanical power' (to use Smeaton's). He could see the *potential* of steam to produce 'work' (our modern term). But what bothered him here was not producing more work, but eliminating waste. Although from the 1850s physicists have been able to talk, in technical language, of different kinds of 'energy' (heat, work, light and so on), in the 1760s there was no such scientific concept and no language of energy. Working in Glasgow College, Watt was concerned not with 'energy' but with an ineffective use of steam.

Some potential sources of waste were clear. They depended on the way in which different substances conducted and radiated heat more or less quickly, and how much heat different substances absorbed (or gave out) as they changed temperature. Watt tried to find out how much heat it took to warm his cylinder and, more specifically, the 'quantities of heat' absorbed by wood, iron and copper (in 1763). In Black's terms, these were 'heat capacities'. For example, Black already knew that wood had a lower heat capacity than metals (i.e., it took less heat to raise its temperature). Watt was careful to say that he had discovered isolated facts, not the over-arching theory which Black told him about.

The copper walls of the cylinder in Watt's new model themselves absorbed and gave up heat during the stages of the cycle, as steam entered a cold cylinder, or water entered a hot one full of steam. They also radiated heat, losing it to the atmosphere. With a wooden cylinder it took less heat to bring about the same temperature change, but wood was better at retaining heat than metals (metals received it and gave it out more quickly). This created a different problem. At the stage of the cycle in which a vacuum was required (and was obtained by spraying in cold water), the cylinder walls became a source of heat. Wood was slow to cool down and this made it harder to get a decent vacuum. Desaguliers had suggested using metals with different 'heat capacities' – and especially brass, rather than the usual cast iron, in large engines precisely because it heated and cooled more quickly. Watt would also turn back to metal cylinders.

He now wanted to work out exactly how much cold

water he needed to cool down the hot steam-filled cylinder. It took less to cool a wooden cylinder (because of its lower heat capacity) but wood had been discounted for other reasons. For a metal cylinder, once he had determined its heat capacity, he could simply multiply it by the weight of the cylinder to find exactly the heat it must give up to reach the lower temperature. To calculate the amount of cold water required to cool the steam to the temperature he required, he assumed, at first, that boiling water and steam contained the *same* amount of heat. But to cool the cylinder *and* the steam took far *more* cold water than he expected. When he injected water to cool the cylinder and the steam, the resulting water was far *hotter* than he expected. Somehow, as the steam had condensed, it had given the 'injection' water more heat. This was puzzling and problematic.

There was another problem. In order to get a perfect vacuum in the cylinder, Watt originally thought that he did not need to cool it, and the steam, much below 100 degrees centigrade. After all, at that temperature, steam normally condensed into water. Then he realised that when the pressure in the cylinder was reduced, hot condensation water turned back into steam (water boils at lower temperatures under reduced pressure). If the walls of the cylinder were still hot, that only made boiling more likely and the situation worse. Steam in the cylinder destroyed the vacuum, continued to push against the atmosphere, and thus reduced the engine's power. To try to understand what was going on, Watt did more experiments (probably in 1765) which allowed him to chart the way in which the boiling temperature of water changed at pressures lower than that of the atmosphere. He wanted to find out what (low) temperature he needed to achieve to ensure that the steam in the cylinder really did condense, and thus give the vacuum he required.

He returned to his earlier question: why did it take so much more water than he had anticipated to condense the steam and to cool down the steam and the cylinder, so that the resulting water was no longer piping hot? When Watt mixed one 'part' of boiling water with 30 of cold water he found it 'scarcely sensibly heated to the finger'. But when he piped steam into cold water he could soon make it boil. In one famous experiment, he turned to his trusty tea kettle. He half filled the kettle with water. Then he put one end of a glass tube into its spout and the other into a jug full of cold water (his 'frigeratory'). When he boiled the kettle (as one does) he found that the amount of water in the 'frigeratory' increased a tiny amount, but the water itself began to simmer.

Now, again, he turned to Black, asking him whether steam might contain more heat than water just about to boil. After all, his own experiments, apparently carried out with his friend James Lind and relying on the balances and thermometers which he had close to hand, suggested that steam from one volume of water could raise the temperature of six equal volumes of water from room temperature to boiling point. Once again, Black had an explanation at hand, and told Watt about his principle of 'latent heat' (the heat required for a change of state without change of temperature). Unknowingly, Watt had measured the 'latent heat' of steam and he had shown it to be high. Contrary to some accounts, Watt did not make an independent 'discovery' of the doctrine of latent heat, nor did he claim to have done so. He was careful to say that the theory originated with Black. With conventional modesty, he later claimed that he had merely 'stumbled upon one of the material facts by which that beautiful theory is supported'.

Understanding latent heat, heat capacity, and losses of heat through conduction in a concrete way through experiments, and discussions with Robison and Black, helped Watt to understand the engine's processes. Having both observed and measured steam's latent heat, he could now understand why the injection water remained so hot. He could see why it took so much extra fuel and fire to push boiling water over the brink to make steam. He could also understand how a small quantity of steam could raise the temperature of the cylinder. The phenomenon of latent heat – if not Black's full-blown doctrine – helped him to track what was going on in the engine as water turned to steam and back again. It is important to recognise that this understanding gave him a far better sense of where the use of steam was inevitable and where steam was being wasted.

Moreover, it applied as well to a colliery engine as to a classroom model. If originally he had wondered why a model worked less well than a real engine, now the question had been subtly changed. How, in practice, could he save steam in a working Newcomen engine, so as to approach the goal of the perfect engine?

The Route to the Separate Condenser: Watt's Perfectible Engine

In fact, the first fruit of all this hard-won wisdom was the certain knowledge that the perfect engine was beyond his immediate reach. In the abstract, at least, heating up a cold cylinder (of known heat capacity) to boiling point itself destroyed a quantity of steam (whose latent heat he had found). Only then could steam fill the cylinder. To make the perfect vacuum, the injection water then had to cool the cylinder walls right down (low enough to ensure that steam could not form to create a back pressure against the atmosphere). He could calculate the minimum amount of steam used in each stroke, but the result was disappointing. He would always need more than one cylinder-full of steam if the engine worked in this way.

Watt later described the dilemma:

That to make a perfect steam-engine, it was necessary that the cylinder should always be as hot as the steam which entered it, and that the steam should be cooled

down below 100° in order to exert its full powers. The
gain by such construction would be double: – first, no
steam would be condensed on entering the cylinder;
and secondly, the power exerted would be greater as
the steam was more cooled. The postulata, *however,*
seemed to him incompatible, and he continued to
grope in the dark, misled by many an ignis fatuus.

The problems were now clear. In an actual engine, by
the time the cylinder was full of steam, it was at least
as hot as boiling water. Cold water then had to be
injected not simply to condense the steam (at normal
boiling point) but also to bring the walls of the
cylinder (not to mention the piston, the packing, the
snifting valve and so on) to the right (cool) temperature to create the conditions for the partial vacuum
required. When steam was let back into the cylinder
during the 'upstroke' of the piston (as the piston rose
to the top of the cylinder), a large part of the steam
was being 'used up' to heat the cylinder – rather than
simply to 'fill it'. On the one hand, all this cooling
down and heating up again was using extra steam. On
the other hand, it seemed to be necessary for the
engine to work at all – at least in its current form.

Increasingly, Watt focused his attention on the
repeated heating and cooling of the working cylinder
for each stroke, and on the fuel – not just steam –
being wasted in that process. He had located and
calculated a major source of waste, not a minor one.
Now he needed to find a way of avoiding that waste,

without rendering the engine useless. Watt could not have *known* that there was a way of saving steam, economising on fuel, and even increasing the engine's 'pulling power'. For 40 years the Newcomen engine had effectively been on a technological plateau. But he must have been *convinced* that it was true. The contemporary history of the practical arts was even then being re-written, and especially in the *Encyclopaedia* of Diderot and d'Alembert, to emphasise the possibility of 'progress', spurred on by potent science. While men of science frowned upon disruptive, something-for-nothing perpetual motion projects, there was also a bubbling euphoria among the mechanics: how often had mankind reached a 'hard limit' to technological improvement?

It was in 1765 that Watt finally hit upon the 'separate condenser'. Strangely, we do not know exactly when. We do know that the idea was clear to him by 29 April 1765, the day he wrote to James Lind about his experiments with a new model of a 'perfect' engine. There is a popular story, first told by Watt's friend John Hart, that the solution suddenly dawned as he wandered pensively across Glasgow Green one Sunday afternoon. (A boulder with an inscription now marks the very spot.) Watt himself recalled the day:

I was thinking upon the engine at the time and had gone as far as the Herd's house when the idea came into my mind, that as steam was an elastic body it would rush into a vacuum, and if communication was made between the cylinder and an exhausted vessel, it would rush into it, and might be there condensed without cooling the cylinder. I then saw that I must get quit of the condensed steam and injection water, if I used a jet as in Newcomen's engine. Two ways of doing this occurred to me. First the water might be run off by a descending pipe … and any air might be extracted by a small pump; the second was to make the pump large enough to extract both water and air … I had not walked further than the Golf-house when the whole thing was arranged in my mind.

Watt's idea was brilliant in its simplicity. First, he would ensure that the working cylinder was permanently as hot as the steam from the boiler. There was to be no injection of cold water there, so steam would never be wasted in heating up the cylinder from cold. Instead, there would be another vessel, connected to the first by a pipe with a valve, but otherwise insulated from it. Watt would condense the steam from the working cylinder in this second vessel, ever afterwards known as the 'separate condenser'. That vessel would always be kept cold so that a (near) perfect vacuum could be created inside it. There was one complication. To draw the steam from the working cylinder into the separate condenser he would need an air pump. That would create a vacuum, sucking the steam in. Here, at least on the face of it,

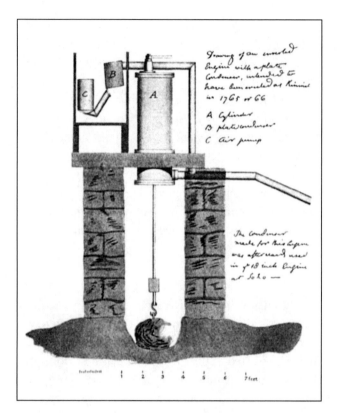

Illustration 7. Sketch of an experimental engine, with Watt's separate condenser, designed in 1765. The intention was to erect it at Kinneil, in collaboration with Watt's partner John Roebuck. This 'inverted' engine does away with Newcomen's heavy rocking beam and lifts directly. The (hot) cylinder (A) is clearly shown, as are two key innovations: the (cold) separate condenser (B), using metal plates instead of Newcomen's spray of water to condense the steam, and the air-pump (C). (Source: J.P. Muirhead, *Origin and progress of the mechanical inventions of James Watt* (1854), vol. iii, plate 1.)

was that elusive engine which used no more than a cylinder of steam for each stroke.

On the Sabbath, Watt was not about to set to work. But on Monday he began to put together a model which, so far as we can tell, used a brass anatomical syringe for a cylinder, and a condenser made of tin-plate and solder and closed with a thimble. The malevolence of things has often manifested itself in undreamed of, but insurmountable, practical difficulties. This time, it took Watt only a few more days to shunt the principle of the separate condenser from the realm of fantasy to a working embodiment in metal and wood. Almost immediately, Watt made a grudging confession to his friend Robison. 'I came into Mr Watt's parlour without ceremony', Robison later recalled,

and found him sitting before the fire, having lying on his knee a little tin cistern, which he was looking at. I entered into conversation on what we had been speaking of at our last meeting – something about steam. All the while, Mr Watt kept looking at the fire, and laid down the cistern at the foot of his chair. At last he looked at me, and said briskly, 'You need not fash yourself any more about that, man; I have now got an engine that shall not waste a particle of steam. It shall be all boiling hot; – aye, and hot water injected if I please.' So saying, Mr Watt looked with compla-cency at the little thing at his feet, and, seeing that I observed him, he shoved it away under a table with his foot. I put a question to him about the nature of his contrivance. He answered me rather dryly. I did not press him to a further explanation at that time, knowing that I had offended him a few days before by blabbing a pretty contrivance which he had hit on for turning the cocks of the engine …

Robison's account makes it clear that Watt believed he had it all sewn up. As soon as you hear the idea of the separate condenser, the solution seems simple and deceptively obvious. But in 1765 Newcomen engines had been working well for half a century and it was hard, in those circumstances, to think beyond them. Who, after all, would bother to re-invent the wheel? It was seven years since Robison had suggested the steam-driven carriage that set Watt thinking. It was perhaps five years since Anderson had set about retrieving the sickly Newcomen model. Watt had been tinkering with steam-filled syringes for three or four years, and experimenting more seriously for two. But from his first year in the College he had seen little point in *ad hoc* arrangements with the science professors, and right through this period he was expanding his business and gradually stepping further and further outside the College bounds. Although Watt had come up with an idea with earth-shattering consequences in engine design, his task in the winter of 1763–4 had not been to improve the Newcomen engine, but rather to fix that irksome model – whose main fault was that it worked less well than its trusty commercial counterparts. Watt really had overshot the mark quite considerably. But it was not without years of experimentation, thought and conversation in unique and fortuitous circumstances. If there was a 'flash of inspiration', it was advertised well in advance.

There is still controversy among those who want to see the beacon of science guiding Watt towards the idea of the separate condenser. Years later, the natural philosophy professor Robison claimed that it was the chemistry professor Black who, illuminating Watt as to the high latent heat of steam, directed him to see the 'economy of steam' as the most important factor in the engine's (re-)design. Black explicitly tried to 'remind' Watt in 1780 that he had known about the doctrine of latent heat early on, that he had been 'directed' by it in his quest to improve the steam engine, and that he had been incited to carry out experiments in connection with it. Watt down-played Robison and Black's claims, but the story persists. Watt also rubbished the suggestion, already gaining currency in his lifetime, that he had been Black's student, pointing out – quite plausibly – that he had been far too busy to attend any of the chemist's classes, however valuable they might have been. In addition, he denied the suggestion that Black had guided him closely in his own experiments with steam.

On physical terms alone, it makes little sense. If the latent heat was lower, more steam would have been required to heat up the cooled cylinder (so the waste might have appeared greater). If it was higher, less steam would have been used up (suggesting that waste of steam was not a problem). From this narrow perspective, at least, latent heat had little to do with it. But we should remember that Watt was lucky to be in Black's company. From him, Watt admitted, he gleaned knowledge, borrowed modes of reasoning, and copied ways of experimenting. Watt's experiments focused on phenomena (latent heat, heat capacity) about which Black was optimally informed. Black may even have funded some of the early experiments, indirectly, since he lent Watt £150 in 1763, no doubt thinking him a safer bet than the banks that crashed, losing him a small fortune. Certainly, Watt had privileged access to a man engaged in parallel experimental programmes, thinking deeply about the properties of heat, caring passionately – and professionally – about reforming the arts and manufactures of Scotland, and prepared to back words with cash.

The Morning After: Watt's Temperamental Engine

So began the long and tangled process of turning ideas about steam, heat and power and *ad hoc* practical arrangements with tin cans and syringes into a working engine. First, there would be a small 'demonstration' model, a little larger than the

College's Newcomen engine. Then there would be an 'experimental' engine, endlessly modified and re-moulded until it was good enough to win over financiers and customers. Finally, there would be literally hundreds of large-scale engines working in factories in Britain and beyond. All this would take Watt and his allies – craftsmen, men of science, financial backers, canny industrialists, lawyers and sympathetic politicians – nearly twenty years of intense activity.

Although the basic idea of Watt's engine might seem straightforward, and its waste-saving potential clear, in practice many additional features were needed, and some of them consumed steam in entirely new ways. As an 'improvement' on the Newcomen engine, the new model was both like and unlike it. Much of the existing technological apparatus, including pumps, housing and boilers, could be borrowed; some mechanisms became irrelevant; and even where it was not absolutely essential, Watt took the opportunity to innovate.

For years, Watt had been preoccupied with steam being wasted by the radiation of heat from the cylinder. To cut this source of waste, in the new model the cylinder would be 'wrapped' in another, insulated with wood. Another source of waste was friction, so Watt used oil as a lubricant to reduce it. To keep every-thing inside permanently hot, the space between the inner and outer cylinders would be filled with fresh steam all the time, to make a kind of 'steam jacket'.

Watt scrapped the 'water-packing' (a pool of water) on top of the piston which made Newcomen's cylinder air-tight as the atmosphere pushed, from outside and above the piston, to develop power. In the new engine, Watt capped the working cylinder at both ends, but to allow the piston-rod to pass through the top and to move up and down without the steam gushing out, he had to develop a steam-tight 'stuffing box' which could also endure corrosively high temperatures. All of this would push contemporary technology to the limit.

Watt's engine really was a *steam* engine', in the sense that steam, rather than the atmosphere, did most of the work. So far so good. In practice, even the simplest Watt engine needed a complex system of valves to regulate the flow of steam. From the beginning, in his very first model, Watt decided that steam should enter the cylinder above the piston (unlike the Newcomen engine, in which it had come in below). The steam drove the piston to the bottom of the cylinder (unlike the Newcomen engine, in which the atmosphere had done the work). The valve letting the steam in was then closed. Another valve allowed the steam to flow freely from above the piston into the space below it. The pressures above and below were now equal, and the engine was balanced, or weighted, in such a way that the piston rose naturally back to the top of the cylinder. Now, a new supply of steam was allowed to enter from above, just as an exhaust valve let the steam under the piston escape to the separate condenser.

In the earliest models, the separate condenser was made out of thin metal and was also immersed in water as a reliable way of keeping things cool. When steam hit the cold surfaces of the vessel, it condensed (so the arrangement went under the name of 'surface condenser'). Watt also tried condensing the steam by bringing it into contact with a series of metal pipes kept permanently cold by flowing water, but neither method could do the job quickly. So, somewhat reluctantly, Watt returned to Newcomen's method: a jet of water.

Next, he had to find a way of removing the condensed steam and, just as importantly, the air brought into the condenser with the steam. Since the air did not condense at these temperatures, there was a danger that it would clog up the system, reducing the 'goodness' of the vacuum against which the piston worked. Watt needed an air-pump (a far cry from Newcomen's simple 'snifting valve') to suck out the polluting air and the tepid condensed steam mixed with its spray of water, to create a vacuum into which the 'used' steam of the working cylinder would rush.

These features sapped steam and made the engine more complex, less tolerant of small defects, and more costly – despite the original intention to improve the Newcomen engine and save steam. It was like the wrinkle in the carpet. The jacket was necessary to keep the cylinder warm so that no steam was wasted in heating it; but it needed its own constant steam supply, over and above that which went into the working cylinder. Watt saved steam, but only by spending it elsewhere. The air-pump, too, was effectively steam-driven. It siphoned off power from the great pump to which the engine was connected (with a huge wooden beam) and which was its *raison d'être*. The crucial question was whether the savings of steam were greater than the newly introduced expenditures of steam. In fact, they were. But mushrooming mechanical complexities still entailed considerable technical difficulties. In most cases, they could be overcome with thought, practical experiment, and special – but therefore expensive – materials and skills. In every case, Watt and his assistants needed to solve a problem, slide around it, or, if they could, jettison a desirable but inessential scheme (like the surface condenser) when it proved intractable. There was still much to be done to finish this fine, upmarket but temperamental debutante for big business.

· CHAPTER IV ·

AN EXPERIMENT IN ENGINEERING

Roebuck's Uncertainty: the Clandestine Engine

Watt alone did not have the resources needed to develop that simple idea, 'the whole thing arranged in my mind', into a viable competitor with the Newcomen engine, entrenched in its considerable home territory. Good family and fine philosophical friends he may have had, but he was no Tobacco Lord ready to underwrite experiments in industry. He had a wife to support, there would soon be children, and he had to earn a living. As a busy but not affluent instrument-maker, merchant, and small-time investor, he could pursue his invention when time allowed. Constructing a crude demonstration model was one thing. But Watt knew he could expect his share of 'intermediate failures' as he built the 'steps to enable me to climb to the top of the ladder'. Watt could not devote all his energies to a machine which might never work. The journey of invention did not start with a solitary 'flash of inspiration' on Glasgow

69

Green, nor did it end there. Without external backers buying time, materials and influence, Watt's engine would have been dead in the water.

He might have turned to John Craig, but in December of 1765, Watt's *annus mirabilis*, Craig met an untimely death. It could hardly have come at a worse time. In their business, Watt had been the manager and the ideas man. Craig had provided the cash, and now his trustees wanted it back, to the tune of £757. His livelihood in jeopardy, Watt cast around for ways to stave off a crisis – even as the promise of the new engine hit home. There must be an alternative backer. Joseph Black was a big-hitter in chemistry, but that did not make him an 'ivory tower' philosopher. In 1759 Watt had gone into partnership with Black and Wilson, selling his skills for hard cash. Black was in demand for expert advice on new industrial processes, from glass-making to vinegar manufacture. He had invested in Watt in 1763, but that money was not to be returned for years. A man with a good nose for promising projects, in the summer of 1765 Black had already urged Watt to contact his former pupil John Roebuck (1718–94).

Roebuck had money, technical resources and a quandary. In 1749 he had begun to manufacture sulphuric acid using his own 'lead-chamber' process, and with the rapid expansion of the Scottish linen industry it was much in demand as a bleaching agent. Roebuck had also founded the Carron foundry, Scotland's first heavy engineering works. His quandary was that he wanted coal from mines that he leased at Kinneil, near Bo'ness, but the enterprise was getting into deep water. He needed pumps to make the seams workable, but even though he was installing the traditional Newcomen engines elsewhere, he did not believe they would be up to the job at Kinneil. Did Watt have the solution?

Watt's engines, then in their infancy, crawling and stumbling in flimsy forms no more useful than Hero's toys, might soon provide Roebuck with the technological fix he craved. Black thought so, and he was convinced that Roebuck, a man with a good track record as a speculative innovator, could be persuaded to finance the engine's development. Black was right. Roebuck saw that Watt's ideas might be realised in a workable engine. He took over the debt to Craig's trustees so that Watt's instrument and merchandising concerns could continue. He offered him the opportunity for engine experiments at Carron, with its facilities unique in Scotland. And he gave Watt the run of the workshop attached to Kinneil House (his own home) for steam-engine experiments on the sly, to be done with John Gardner, Watt's most senior assistant and one-time apprentice. It was still up to Watt to convince Roebuck to invest serious money, and that meant making a working experimental engine.

But when progress was slow, Watt wavered. In 1766, Black decamped to Edinburgh. Craig's death had not destroyed Watt's business, but it marked the beginning of the end for his little empire. Between 1771 and 1773 he gave up the instrument business, hiving off the philosophical trade to Gardner. There was little time for invention. James Watt, workshop manager, business-man and merchant, was transforming himself into James Watt, engineer. It was not a complete surprise. Watt's uncle John Watt had worked as an engineer and surveyor before his death in 1737. Watt sketched surveys in his childhood sketch-books. As an instru-ment-maker, he sold theodolites and other surveying tools. And in 1758, Watt had dabbled in surveying work alongside the versatile Alexander Wilson.

For this occupation, Watt had both role model and competitor in John Smeaton. Another middle-class boy, Smeaton too trained as an instrument-maker at a time when civil engineering was not yet an independent profession and there was no established training for it. Even Thomas Telford, first President of the Institution of Civil Engineers (founded in 1818), started out as a stone-mason. Fêted as an inventor and mechanic, Smeaton was the intellectual beacon behind the new Eddystone Lighthouse of 1759. Yet as the archetype of the British civil engineer, Smeaton did more than anyone else to promote its claims to be a paid-up – and well paid – profession, distinct from military engineering and the traditional crafts. It was not only, then, because of Smeaton's deep investigations of real, and perfect, waterwheels that Watt looked upon him as a model of 'perspicuity of expression & good sense'.

From 1766, engineering took up much of Watt's time. He was no specialist, at a time when few engineers could afford to be. Like Roebuck and Black, he entered the cauldron of industrial chemistry. He gave technical advice to the cracked and flaky Delftfield Pottery company. In 1766, he delved into the murk of the mining business. But in these days the best money flowed quickly and steadily from the canals which were just then opening up the industrial heartlands of Britain. James Watt, Engineer, wanted to be part of it.

His first major job came in 1767 when he joined Robert Mackell to fix a route, via Loch Lomond, for a canal linking the Forth and Clyde rivers, uniting east and west Scotland and, most importantly, keeping the Tobacco Lords happy. Outrageously, Smeaton suggested a route bypassing the city. Watt and Mackell obliged with a terminus in Glasgow itself. For a few heady years, and to his surprise, Watt had all the work he wanted. He played up his local credentials and undercut more experienced engineers, including Smeaton, to win contracts. Success brought rewards, but it also brought a dilemma. In 1769, Watt admitted: 'I would not have meddled with it [i.e., the canal work] had I been certain of being able to bring the engine to bear; but I cannot, on an uncertainty, refuse every piece of business that offers.'

If canal work took Watt away from the engine, the experience of steam engineering was absolutely invaluable. Watt hacked out a niche for himself designing, constructing and maintaining Newcomen engines. Again he worked with, and learnt from, Mackell, who knew the atmospheric engine business well. The Carron connection was vital and timely, since in 1765 the company was just attempting to break into the markets for Newcomen engines, especially in Cornwall but also in Scotland. Watt helped to set up a small Newcomen engine at Carron in 1765, and he built others with Mackell in 1766 and 1767. An atmospheric engine at Ayr, finished in 1770, was so exceedingly splendid that it threatened to send its new owner 'mad with joy'. Watt knew exactly what he was about. He knew that he lacked 'the necessary experience in great'. No amount of reading, model-making and experimenting in Glasgow College could compete with hands-on experience in the guts of the best atmospheric engines of the day. Once he had that experience, he told Black, he was 'not the same person' who thought up the separate condenser in 1765.

Thus armed, Watt went back to work on the engine in January 1768. Trials went ahead, but it wasn't plain sailing. In April 1768, an accident left mercury from a gauge splattered over the cylinder's insides and Watt, much vexed at wasted time, spent three days getting things in order. This was just one of many tribulations. By May, Watt managed to put together a small model 'test engine' which served its purpose in persuading Roebuck to back him, but at a price. Roebuck promised to stump up the cash for Watt's development work, and agreed to pay to patent the engine, in return for 'two thirds of the property of the inventions'. That meant a hefty two-thirds share in the future profits of the engine and anything arising from the patent.

Up until now, Watt had relied heavily on allies in and around Glasgow, but even with Roebuck's support, it was time to cash in on friends further afield. Travel was a vital stratagem in Watt's campaign to complete the engine. His first sojourn in London, although profitable and even revelatory, had ended in a sickly retreat to Scotland. But Watt returned to London several times to stock his 'toy shop', and contemplated trips to 'Leverpool' to flog his nautical instruments. He met with élite London instrument-makers like Jesse Ramsden, and exchanged useful professional gossip. These journeys were oppor-tunities to gen up on local industries. In 1767 he visited Birmingham, not for the first time, and there met Roebuck's friend, the medic Dr William Small (1734–75). In his reincarnation as a natural philosopher, Small shared Watt's fascination with steam engines. It was in July 1768 that Watt trudged to London again to start the tortuous procedure of applying for a patent. On his way back to Glasgow, he passed through Birmingham once more, and there Small was waiting.

The fruits of Watt's friendship with Small were little short of spectacular, not least because of Small's own circle of friends. In Watt's *annus mirabilis* of 1765,

Small had been given an introduction to one Matthew Boulton (1728–1809) by an American acquaintance, the statesman Benjamin Franklin. Small was soon Boulton's physician. As an entrepreneur, Boulton outdid Roebuck in his thirst for the new. He exuded dynamism, courted innovators, hobnobbed with the rich and famous, and craved success. Boulton had inherited a buckle business from his father and had begun to make money by turning out elaborate shoe ornaments and silverware from a town (Birmingham) dominated by those industries. Small introduced Boulton to Watt and they hit it off immediately. Watt was Boulton's personal guest for an entire fortnight.

Although Watt had had a sneak preview of Boulton's huge Soho works in Birmingham with Small in 1767, he now had a guided tour from the owner himself. This busy establishment, opened as recently as 1762, drew technological tourists from all over Europe. They marvelled at its fine commodities, from trinkets to clocks, produced with state-of-the-art equipment in an ostentatiously progressive environment, and on the latest principles of factory organisation. Boulton employed 600 at Soho, but wherever he could he replaced human labour and skilled craftsmanship with power-driven machines. Watt could only look on with envy. He knew already that Soho offered what he needed, and what Roebuck lacked, to finish the steam engine.

Despite Boulton's openness to innovation, he was short of options when it came to power. When the nearby Hockley Brook was in full flow, a mill wheel provided the power he needed. During summer droughts and winter frosts, the flow was sluggish. Boulton, like other factory-owners, had to buy in horses to drive the pumps to raise the water to work the wheel. In his modern factory, predicated on continuous production, he wanted more power to meet growing demand for his quality products, and he wanted a constant supply, unconstrained by the unpredictability of the British climate or cycles of seasonal change. Looking for something better, industrial entrepreneurs turned to the best available 'horseless' pump. The coal-guzzling Newcomen engine was an expensive option, but Boulton toyed with the idea of getting one in 1766 to pump water back and over his mill wheel. By then, of course, Watt knew first-hand about the foibles of these machines.

When Boulton first heard about Watt's 'improvements' from Roebuck and Small, he had been intrigued but had decided to hold fire. Boulton's penchant for the untried and untested had strained his bank balance before, and investing in Watt was a gamble. But with Watt picking up the threads of invention so well in the summer of 1768, and especially producing a working model for Roebuck at Carron, Boulton was primed. In Birmingham, he grilled Watt on steam's prospects, quizzed him on the secrets of his invention, and offered him help. Watt was still bound to work with Roebuck, but he wanted Boulton in. Roebuck responded in October 1768 by offering Boulton the right to manufacture engines under a patent, soon to be obtained, but only in the three nearby counties of Derbyshire, Staffordshire and Warwickshire. In December 1768, he teased him again with a share in Watt's 'New Plan of an Engine'. Boulton was not interested in half measures. Famously, he told Watt early the next year that he preferred 'to make for all the world'.

Lessening the Consumption of Steam: Patenting Principles

There remained the matter of securing the invention lest other inventors took the idea, or came to it independently. It was on 5 January 1769 that Watt obtained a patent for his 'fire engine' with its brilliant novelty: the separate condenser. That was nearly four years since that fruitful Sunday stroll on Glasgow Green. In the meantime, Watt had several times almost lost faith in the engine. Short of money, he was drifting towards civil engineering and the steadier income it offered. Even with Roebuck's support, invention was a burdensome business and patents were expensive beasts entailing lengthy legal manoeuvres. Watt needed allies to buy protection for his ideas.

The title of the patent as a 'New Method of Lessening the Consumption of Steam and Fuel in Fire Engines' gave Watt's aims away. The Newcomen 'fire engine' consumed fuel, gobbled up steam, sniffed away air, and dribbled tepid water. Watt believed that he had discovered the principles of action which could sort out this bad behaviour. Watt was expected, at least by custom, to present the High Court of Chancery with a 'specification' or a description of his invention detailed enough, in theory, to allow another engineer to make one, once the patent expired. It was always easier to copy objects, which is what Watt had done with quadrants and compass boxes, or to sneak a look at drawings, or chat with the constructors themselves. In practice, then, even the best written accounts were insufficient for all but the most experienced engineers to replicate an invention.

In fact, Watt found it extremely difficult to decide how to frame a technical description of his nascent invention. He scrimped on legal advice, which in any case was conflicting, and turned to scientific friends, who wanted credit only in the form of recognition. For witnesses he turned not to lawyers but to George Jardine, Roebuck's friend and eventually Glasgow College's logic professor. Although Watt prepared a description and diagrams of a workable engine, he did not, in the end, include them. Both Small and Boulton counselled Watt not to patent a specific machine or application but rather his method, presented as a list of guiding principles. Much legal ink would be spilled in the 1790s over whether a 'method' alone could be patented. But in 1769 the patent framed in this way staked out a wide territory for

Watt, prevented him from being outflanked by another inventor coming up with a specific mechanical alternative, and labelled more mechanical imitators as pirates. It greedily captured practically every *possible* steam engine with separate condenser. Later, steam innovators would fume with resentment when, marked as infringers, they were hauled before the courts.

It was an earth-shattering document. Back in 1769 the patent did, at least, outline the main features of the engine that Watt had conceived in 1765, with some subsequent additions. Here were the now-familiar jacket to keep the cylinder steam-hot, the separate condenser, the air pump, and minor details like oil lubrication – all designed to lessen the consumption of steam in fire engines designed primarily for use as pumps. But scattered within the patent were the seeds of more adventurous plans. The Newcomen engine was pretty much a one-trick pony. The down and up movement of the piston transmitted in a see-saw action by a heavy beam, translated into an up and down movement which worked a pump. Back in 1766, Watt had been thinking about how he might make his steam engine more than a glorified pump. Smeaton's wind and waterwheels already produced rotary (circular) motion and deployed it in a multitude of secondary machines vital to industry. To compete with them, Watt needed to harness the great natural force of steam and get rotary motion directly. Insuring himself against the future, Watt included a hazy description of a 'rotative steam engine' as the 'fifth principle' of his patent of 1769. Wonderfully simple in design, it proved impossible to build. Would that be the fate of Watt's other ideas?

Partner Swapping: From Roebuck to Boulton

In return for his share in the patent, Roebuck had agreed to foot the bill for developing an engine with the new features of the separate condenser and air pump on a 'working scale' at Kinneil. This engine, constructed in the spring and summer of 1769, was an order of magnitude bigger than Watt's little Newcomen model, or his experimental models with their tiny six-inch cylinders. 'Scaling up' from a miniature engine to an experimental one on a 'working scale' threw up problems, but that was half the point. The engine itself was a test-bed for the new technology, like a laboratory for Watt and his assistants.

Equally, it was a way of demonstrating 'invention at large'. A working engine at Kinneil, once revealed, would be the most tangible evidence to financiers, and eventually to customers, that the Watt engine was more than a pipe-dream. With the Kinneil engine in place, in September 1769 Roebuck made a second offer to Boulton of a one-third share in Watt's patent. Yet managing big experimental trials was nerve-wracking. There were so many reasons to keep quiet about promising new technologies. Even Boulton was under strict instructions to speak to no one except Small. It took Watt until the spring of 1770 to make the separate condenser work properly. It was in these conditions, anxious to publicise, yet desperate to keep things under wraps, that he continued his clandestine 'steam research' in the grounds of Kinneil House.

Civil engineering work was still sapping Watt's energies and eating into his time. He rode away from his experiments at Kinneil to survey the Strathmere canal, near Perth. Roebuck had done his best to bankroll the engine, but his mining interests in Bo'ness proved a bottomless pit for limited reserves of cash at a moment nothing less than crucial for Watt. By the time Watt headed back to Kinneil in September 1770, with the intention of finishing his tests, Roebuck was so financially embarrassed that he had withdrawn from the Carron Foundry. Once again, Watt had a choice: he saw himself working on the engine and 'growing gray' with his wife and children still unprovided for; or embracing the honourable employment of the civil engineer. At that moment, there was no contest. There were plans for a route between Fort William and Inverness, schemes to improve the navigation of the rivers Forth and Devon, and a bright idea to bring cheap coal direct into Glasgow. From 1770, the last of these – the Monklands Canal – netted Watt a steady £200 a year when the steam engine was idle and Roebuck was busy fending off creditors.

In any case, the experimental engine was throwing up intractable problems. It was becoming clear that, in order to contain steam rather than air, it needed a cylinder and a piston tougher and more finely engineered than any found in the Newcomen engines with which, by now, Watt was intimately acquainted. Help might have come in the person of John Smeaton, who, as recently as 1770, had designed a new boring machine and the next year saw it up and running at the Carron works. But the best machine in Scotland was not yet capable of making a cylinder meeting Watt's requirements. This would be a fundamental stumbling block.

To realise the advantages of the steam engine, he needed a good cylinder and specialised mechanical parts. And that wasn't all. In the market place, he would need support systems – from skilled labour to machine-minders, to basic supplies of fuel, water and lubricants. This complex puzzle required a complete solution. Although some answers were ready made – in the culture of the Newcomen engine – others had to be borrowed from further afield or worked out from scratch. History was littered with similar schemes which had never escaped the workshop, or died the death of commercial competition. Success would be astonishing, not inevitable. And now, no cylinder meant no engine.

Watt could find no solace in Roebuck. His patron's business was in dire straits, and in March 1773 he withdrew from the mines around Kinneil, unable to meet his financial obligations. Boulton's waiting game had paid off handsomely. To settle with him, in May 1773 Roebuck signed over his two-thirds share in the steam-engine patent. Watt discharged Roebuck of his obligations in return for the experimental engine at Kinneil. 'None of his creditors', he told Boulton, 'value the engine at a farthing'. Watt packed it up and shipped it to Soho where it remained, in fragments, since he was too busy to travel south to re-assemble it. In September 1773, his wife died in childbirth, leaving him utterly depressed. He surveyed yet another canal (the Caledonian) over the winter, but even that work was drying up. It is hardly surprising that Boulton coaxed him to make a clean break and try the engine once more.

With the coming of spring, Watt took stock – and decided to take up Boulton's offer and head off for England. He had an ideal travelling companion. Back in the 1760s, Black had introduced Watt to a confidant and co-researcher by the name of James Hutton (1726–97). Hutton too had become embroiled in the management of the path-breaking Forth and Clyde canal. His real enthusiasms, though, were agricultural improvement and geology, subjects which led him to speculate wildly, to question the literal truth of scripture, and famously to see in the earth's history 'no vestige of a beginning, – no prospect of an end'. Moreover, geologising left Hutton saddle-sore: 'Lord pity the arse', he groaned, 'that's clagged to a head that will hunt stones'. But in May

1774, the 'famous fossil philosopher' agreed to travel south with another amateur geologist (James Watt) en route to Birmingham. After years of scheming, Boulton at last had what he wanted.

CHAPTER V

LEARNING INDUSTRY

James Watt, 'Lunatic'

Watt was taking a gamble by entering Boulton's stable. Hutton worried that 'Mr Boltons board is ill qualified for sharpening tools that are best ... forged by the hand of the hungry mother of invention'. Why, shielded by such rich industrialists, should philosophical 'bougres' like Watt get away with causing society's degeneration, turning 'nature upside down' by inventing 'cylinders full of steam with condensers at their arse'? Samson, after all, turned the mill himself, slept with a whore at Gaza, and still rose the next day to carry away the 'gates of the toon'.

In all seriousness, Watt left behind in Scotland the family, academic friends and business contacts who had helped to make the new engine viable. Although Watt was competitive, and, as Robison's anecdote about the separate condenser confirms, extremely jealous of his ideas, he needed allies and, in the right circumstances, he wanted to operate collaboratively. Uprooted from Scotland he certainly was, but compensation came in the form of a concentrated group of industrial sympathisers, the like of which even Glasgow could not offer.

The British loved their clubs and societies as exclusive centres of wheeling and dealing. They were sanctuaries of co-operation, of curiosity, and of commitment. They were places to patronise and to pool knowledge of every variety – from prose to philosophy to practical arts (or technology). London had its Society for the Encouragement of Arts, Commerce and Manufactures (later the Royal Society of Arts), bent since its foundation as recently as 1754 on mankind's improvement. Edinburgh matched it in the same year with a Society for Encouraging Arts, Sciences, Manufactures and Agriculture. The Scots offered cash prizes to entice industrial innovators to share their spoils for the nation's benefit. Rivalling London and Edinburgh, assertive, confident and enduring provincial groups – like the Manchester Literary and Philosophical Society of 1781 – brought together industrialists, doctors and 'marginal men', especially religious dissenters who were otherwise denied cultural expression in Britain's centres of power and learning. But before Manchester's 'Lit & Phil' crystallised, Birmingham was already boasting its own quirky 'Lunar Society'.

The Lunar Society first gathered in about 1766. With only fourteen members at its peak, it was small but select, numbering among its 'gigantic philosophers' no fewer than eleven Fellows of the Royal

Society of London. A close-knit and secretive affair, with no constitution and no journal, it matured into a vibrant, if private, meeting place for élite Midlands industrialists. Affluence gave these men time for science at home and for six-hour communal de-briefing sessions at monthly intervals, first on the Sunday and later on the Monday nearest the full moon to allow for a safe journey home. It was this arrangement, of course, that gave the society its name.

Members steered clear of contentious political and religious talk, instead devoting their prodigious energies to exploring a potent brew of chemistry, electricity, meteorology, medicine, botany and geology – always with a keen eye on the industrial pay-off. It seemed that no task was beyond them: for nine years, by the clock, at 8 a.m., 2 p.m. and 8 p.m., Watt recorded pressure, temperature, humidity, wind direction and rainfall at Soho in a hopeless attempt to fathom the British weather. With useful science and the practical arts in harmony, however, the society of 'fellow schemers' looked forward to enhancing the economy, material conditions and international standing of Britain – and in the meantime, doing themselves some good too.

Chief among the Lunar Society's founders were Boulton and Small, both by that time long-standing allies of Watt. While teaching in Virginia, Small had been at the centre of a coterie of scholars and students, including one Thomas Jefferson. Small was the glue that bound the Birmingham group. After Small's death in 1775, Boulton became the real galvanising force and the group met regularly in his 'Hôtel d'amitié sur Handsworth Heath'. Erasmus Darwin (1731–1802), also in at the beginning, was a physician, poet and naturalist, as well as the grand-father of evolutionary theorist Charles Darwin. Darwin had been with Small when Watt first gazed agog at Boulton's factory back in 1767, and he shared with them an interest in the steam engine. The irrepressible Erasmus gushed: 'what inventions, what wit, what rhetoric, metaphysical, mechanical & pyrotechnical, will be on the wing' at the next society meeting.

The other members, less fun if more solid, shared Watt's interests: in ceramics (as technical consultant to the Delftfield company); in the canal business (as a surveyor in Scotland); in carriages (projecting steam vehicles with Robison); and in chemistry (experi-menting on steam beside Black). The potter Josiah Wedgwood (1730–95), for example, combined artistry, technical experimentation and mass production in products such as his famous Jasperware. Wedgwood's 'Etruria' factory, like Boulton's Soho, drew in hordes of swooning technological tourists. Wedgwood watched materials pour in and finished pots flood out along the new improved waterways like the Grand Trunk Canal, for which he and Darwin had lobbied so hard. Then there was Darwin's 'mechanical friend' Richard Lovell Edgeworth (1744–1817), father of novelist Maria Edgeworth and inventor of an award-winning 'perambulatory land-measuring machine'. His true métier was as a carriage designer, an occupation he copied from Darwin and which earned him the title 'philosopher of transport'.

Most of all, the captains of industry understood the usefulness of chemical science. James Keir (1735–1820) made synthetic alkali and used it to manufacture glass and soap, thereby distilling a healthy profit at his flagship Tipton chemical factory, perched on the banks of the Birmingham Canal. With cash handouts from Boulton, the radical preacher Joseph Priestley (1733–1804) achieved international fame as a chemist, and accumulated one of the best collections of apparatus in the country – only to see his home and laboratory wrecked when reaction to the French Revolution boiled over in the Birmingham riots of 1791. Wedgwood provided Priestley with specialist apparatus. Priestley supplemented Wedgwood's empirical knowledge with advice on clays, ceramic chemistry and coloured glazes. So inspiring was Priestley, spiritually and scientifically, that Keir had to reprimand Boulton for his copy-cat chemical 'hobby-horsicality'.

Between 1765 and 1780, Watt, Black, Roebuck and eventually Keir struggled to find a way of getting alkali from common sea salt and quicklime on an industrial scale. Roebuck hoped to operate the process and it was he who approached that 'mighty chemist' Keir for advice in 1770, long before Watt moved down to Birmingham. Although Watt could make it happen on a small scale, scaling up proved next to impossible and there had been almost no significant progress by 1780, when Alexander Fordyce appeared on the scene and patented a similar technique. Keir thought Fordyce had stolen it from him. Watt labelled him 'a lyer, an Irishman, a Bankrupt & a Blockhead': Fordyce's speculative banking practices had lost Boulton money. Although Watt's scheme flopped, and in 1781 he abandoned it to concentrate on the steam engine, the Lunarmen had been equally supportive of both. Each project illustrates Watt's compulsive determination to use long-term systematic experiments to nurture tiny germs of ideas, or chance observations, into mature industrial processes, whether chemical or mechanical.

Finally, the Lunar Society created machines large and small and engineered them to high standards of precision. Members designed machines for industry: hydraulic rams, presses for stamping coins, and lathes (made at Soho) for turning Wedgwood's pots. John Whitehurst (1713–88), an instrument-maker like Watt, augmented human happiness by re-inventing the water closet. He also designed an accurate assay balance for the new Birmingham Assay Office. Boulton made famously accurate thermometers. Wedgwood dedicated himself to his pyrometers. In sum, the Lunar Society members produced gauges to measure pressure, counters to evaluate engine performance, and clocks to regulate factory workers.

Like Watt, they were clued into the cultures of precision, banishing waste, and standardisation, rationalising and universalising control.

Thanks to the Lunar Society, Watt could also rub shoulders with the leading lights of science and engineering as they passed through Birmingham. These included the naturalist and explorer Joseph Banks, who, from 1778, dominated British science as an autocratic President of the Royal Society of London. (Boulton made beads for Banks so he could barter with new peoples in far-flung lands.) William Herschel built huge telescopes and, with his sister Caroline, scoured the heavens for nebulae – only to find a new planet (Uranus). Small's friend Franklin, not just a statesman but also an electrical adept, added a spark to the Lunatics' meetings. Another visitor, John Smeaton, tried and failed to muscle in on the developing steam-engine partnership but still invited Boulton and Watt to join his new London club (the Smeatonians) designed to bring together the new 'civil' engineers.

Among these men, Watt was in excellent and stimulating company. The Lunar Society was a fertile test-bed for new ideas. Here, Watt could ponder the technical problems of the steam engine destined to be such a boon to Boulton, just as Erasmus Darwin mulled over the horizontal windmill which Wedgwood thought, wrongly, would soon power his pottery. Not everything worked: there was even comical talk from Darwin of an engine, driven by charcoal gas, working on some principle of 'internal combustion'. A sensible man like Watt pooh-poohed such outlandish ideas, turning back to his industrial friends for sage and solid advice on the broader, but necessary, issues of steam-engine marketing and distribution. They showed him how to do something even he had not attempted before: how to make it as an industrialist.

New Life for Old Patents

At the beginning of 1775, Boulton and Watt formalised their previously informal agreement and forged the business partnership of 'Boulton & Watt'. The foundation of that partnership was Watt's patent of 1769. By the time Watt settled in Birmingham in May 1774, the patent was already five years old. By then, a considerable investment of time, money and materials had delivered scant rewards. Only eight years remained in which to recoup the heavy costs of development and marketing. Boulton, and Hutton too, saw that the partnership of 'Boulton & Watt' could hope to turn a profit only if the patent's life could be extended. That was easier said than done.

Extending the patent meant orchestrating a special Act of Parliament. This is precisely what Savery had done back in 1699. Amazingly, it was better to do this than to start from scratch. An Act, at £110, was cheaper than a new patent, at £130, and Watt wanted to save himself £20. He knew, also, that he could rely

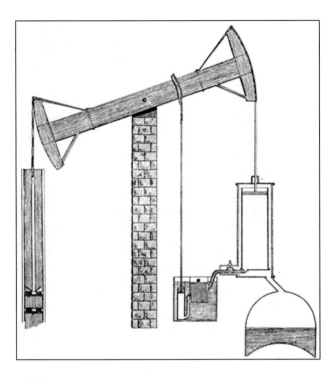

Illustration 8. Watt's engine in the form presented to
Parliament in 1775 as part of the campaign to extend the
patent of 1769. On the right is the cylinder with the
piston working within it, connected by a rod to the (re-
instated) wooden beam. The cylinder is topped by the
'stuffing box', surrounded by the steam jacket, and
supplied with steam from the boiler below and to the
right. Below and to the left of the cylinder, immersed in
cooling water, are the separate (plate) condenser and the
air-pump, the latter being driven by a rod connected, like
the piston, to the beam. On the extreme left is the reason
for it all: the pump. (Source: Doldowlod Papers,
Birmingham Reference Library.)

on Boulton's consummate lobbying skills to get the Act made law. Years previously, Boulton had gone to the heart of government to pontificate on the buckle business and, more recently, he'd won Birmingham its own regional Assay Office. Small jested about Boulton's influential connections: 'I hope the King and royal family, the Nobility and the Ministry and your other friends are well.' Watt, too, had some experience in these matters, gleaned when he was drumming up support for the Forth and Clyde canal in 1767 – although that had hardly endeared him to the London MPs, nor they to him. And Watt now had his own influential friends – like Irvine, who promised to 'pimp' for him in London, or Wedgwood, who offered to swing 'a vote or two', not to mention senior lawyers and engineers, and even the Duke of Buccleuch, a progressive landowner and patron of Adam Smith. This was the payoff for the years spent as a civil engineer.

Watt prepared a pamphlet to initiate a propaganda war, comparing the new engine favourably to Newcomen's, emphasising the hard graft of getting the idea of the (perfect) engine out of his head and perfecting its real counterpart for practice, and giving government two options: pay to make it public; or extend his monopoly. He petitioned Parliament in February 1775, after which a Select Committee, stacked with Boulton's friends, met to consider the matter. The Bill was ready by March but by then opponents were surfacing, lobbing counter-petitions and speaking out for vested interests. Since the 1769 patent had been vaguely framed by amateurs anxious to cast a wide net, it was by no means certain that the Bill would survive. William Blakey claimed, feebly, that the patent infringed his own. Rival engineers railed against the dire consequences of an extended monopoly. The big-hitting politician Edmund Burke, a staunch opponent of American independence, spoke for the mine-owners among his Bristol constituents, who wanted the engine quickly, and for free. Meanwhile, Lord Cochrane was warning Watt that 'Coal Masters, have no Reason to rejoice at any improvement that diminishes the consumption of Fewel'.

Had this opposition proved successful, Boulton would likely have cut his losses. But Boulton had patrons and backers in every corner. He called directly on MPs, especially those hailing from the Midlands who owed him favours. In April 1775, Boulton and Watt gave evidence themselves. Although Watt groaned at the 'great Expence & anxiety', he was soon rejoicing at having surmounted 'a series of various and violent Oppositions' through the support of 'many friends of great intrest'. In May 1775, the patent was extended until 1800 in an Act sanctioned by Royal Assent. A combination of financial resource, legal manipulation, effective networking and Boulton's parliamentary *savoir faire* breathed life into the engine once again.

This was a momentous achievement. Historian Eric

Robinson rates it as perhaps 'the most important event in the Industrial Revolution'. Without it, Watt's engine might never have been 'perfected'. The 'steam revolution' at the heart of the revolution in industry would surely have been delayed. Yet in the years to come, opponents of the extension would look on as Boulton and Watt deftly played the patent card to out-trump rival steam engineers – even, perhaps especially, those who promoted more efficient engines. Extending a monopoly might be good for Thomas Savery, or for Boulton and Watt, but it was disastrous for their opponents, stifling competition – and, in the terms of the day, hampering or altogether blocking technological progress at a time of huge economic growth and innovation elsewhere. It was no accident that the most fuel-efficient high-pressure engines of the early nineteenth century were developed in France.

That was in the future. Now, with the Act in place, the project moved forward rapidly. Thanks to the engine, Hutton claimed, 'Mr Bolton [*sic*] will never want water in the driest summer' for his power mills. Boulton could now throw all his weight behind the project without fear that the potential benefits would go to others. But to market their machine, Boulton and Watt needed a full-scale working engine, at first in the safe haven of the Soho works, where it could be shown off, under their control, to best advantage. Better than any small model, or unconvincing scientific explanation, a working machine would provide managers with proof positive of the steam engine's merits.

This was assuming, of course, that it could be made to work at all. Watt had had to re-invent himself as a steam engineer to develop the engine and promote himself as its chaperon. But as early as February 1769, Boulton had recognised that success and reputation would come only if Watt's engine were constructed with 'as great a difference of accuracy as there is between the blacksmith and the mathematical instrument maker'. They wanted vast engines with philosophical precision. They needed an accurately engineered, steam-tight working cylinder, but all they had was Roebuck's inadequate casting.

Exciting changes in the machine-tools industry offered hope. Boulton pointed Watt in the direction of iron-master John Wilkinson (1728–1808), Joseph Priestley's brother-in-law. At his base in Bersham, North Wales, Wilkinson had a new boring mill for cannon-making. He now used it to make a cylinder with an accuracy far exceeding anything Smeaton could do. Eventually, Watt would be agog at the way Wilkinson could make a 72-inch cylinder with deviations of less than the thickness of a 'thin sixpence'. That meant errors of only one in a thousand. In April 1775, as MPs deliberated over the patent's life, Watt was not yet so impressed, finding Wilkinson's work 'not perfect, yet there doth not appear any very gross error'. It was good enough. Into the new cylinder slid a finely machined piston, 'packed' after many failed trials with hemp soaked in tallow to make a near-perfect seal. State-of-the-art machinery and 'make do and mend' combined, preparing Watt's steam engine to leave Soho and contend in the theatre of power.

A County of Fire Engines

It was not until 1776, a decade since the conception of the separate condenser as a route to the perfect engine, that practical Watt engines finally made their mark on industry. The processes of marketing and making were anything but automatic. Political economists, mine-owners and industrialists needed to be convinced to take a chance on risky innovations. In his *Wealth of Nations*, published in 1776, Adam Smith said nothing about the new steam engines. As for the industrialists, they could easily guess that Watt's engine was more complex than Newcomen's, and they could not know that fuel savings would outweigh construction and maintenance costs.

Faced with that knowledge, Boulton and Watt chose their first customers carefully. By minimising the unknown, they gave novelty a chance to flourish. They turned to men they knew, or to nearby places. One early engine went to Wilkinson and his blast furnaces at New Willey in Shropshire. The first pumping engine in Scotland was for Peter Colville, familiar to Watt as a customer for his own atmospheric engines. Going further afield was risky. Smeaton had painstakingly doubled, but only doubled, the efficiency of the common Newcomen engine by 1772; called upon to inspect the first Watt engine in London in 1777, it was rumoured that he took the opportunity to douse the engineer in drink. Small wonder that the next day it was 'almost broke to pieces'. The very first engine to be completed, in March 1776 and prematurely in Watt's view, served the Bloomfield Colliery, near Keir's Tipton works. That engine made its first tentative pump amid tremendous pomp at a grand ceremony attended by curious scientific gentlemen and stage-managed to inaugurate the new era of Boulton & Watt. Claims that 'the Doubts of the Inexperienced are dispelled, and the Importance and Usefulness of the Invention is finally decided' were certainly premature, but this first showcase really was good for business. The Cornish engineers were some of the most enthusiastic visitors.

The main target destination for the engines was, of course, Cornwall with its tin and copper mines. Two things made these deep mines viable: high prices and good pumps. Since the 1710s, Newcomen pumps had proved themselves the real 'miner's friends', and the county had 60 of them by 1775. Even then, they consumed coal at a great rate, and coal was expensive in Cornwall because it came in from a distance by sea. Cornwall was the most gruelling testing-ground for any pumping engine. Mine-owners kept detailed records of performance, and any engine promising greater efficiency was of particular interest. In these first years of commercial exploitation, it was the place where Boulton & Watt's fuel-efficient engines would find most favour. In 1776, Boulton and Watt took orders for two engines, one of which went to Tingtang, near Redruth, and the other – not completed until September 1777 – to Wheal Busy. By 1800, there were 55 Watt engines in Cornwall. Newcomen had been sidelined.

Instead of waiting for business to come to them, Boulton and Watt invested time away from Soho in promoting the engine, convincing potential purchasers that this was a better buy than existing prime-movers. Boulton looked after the finances, while Watt worried about design and construction. Between 1777 and 1781, the partners often left Keir in charge in Birmingham. Keir doubled as money-lender, keeping the business afloat, and management consultant. In 1778, Boulton & Watt had debts of £42,500 and 'total incomings' of only £4,000; so with the best will in the world, Keir could only conclude that the company's finances were parlous and its prospects dismal. Watt was so often in Cornwall that he set up home there, even though he loathed the county, found the Cornish uncouth, was convinced (as usual) that his health was suffering, and hated the management of workmen and the face-to-face bargaining that was necessary to bootstrap Boulton & Watt into profit. 'Nothing is more contrary to my disposition', he had once told Small, 'than bustling &

bargaining with mankind'. There was something else he had learnt working on the canals: 'I would rather face a loaded cannon than settle an account or make a bargain.'

Watt was not cut out for this coy technological courtship, with its inauspicious squabbling over payments and parts, contracts and delivery dates. When this old technology was new, there were no 'off the peg' engines, set up in moments and left to run in perfect order. The truly perfect steam engine receded as Watt junked the steam jacket and jettisoned the surface condenser. But he still wanted engines that would make and maintain the company's reputation at the highest level, assembled with philosophical precision rather than cobbled together. There was no team of trained experts to call in. Watt said that he had 'not only to form my own Engines but also my own Engineers'.

Most especially, he looked to William Murdock (1754–1819), a pattern-maker at Soho from 1777. Having watched the engines mature around him, it was small wonder that Murdock had a special, perhaps unique, knack for on-site construction and trouble-shooting repairs. Local men would work beside Watt's highly skilled engine 'erectors', learning from them how to make the engines, how to keep them running, and how to optimise their performance. Only in 1779 did Watt get around to composing a lengthy set of *Directions for erecting and working the newly-invented steam engines*, and even then it was hard to get it right. Some things could not be written down. Others were obvious, and jokers supplemented this manual, and later ones, with hilarious hints like 'Don't leave your hat, coat or shoes in the cylinder'. In this context, Murdock was irreplaceable. Spending most of his time in Cornwall between 1779 and 1794, he saw to it that the revolutionary new pumping engines became reliable and unproblematic features of the industrial landscape.

The mine-owners paid Boulton & Watt a 'royalty' on their patent. It was calculated in an unusual way. In pumping water, the more efficient Watt engine consumed less fuel than the common Newcomen engine. The idea was to tally up the amount of fuel saved against this standard (but not against one of Smeaton's enhanced engines). This led to a notional cash saving for the mine-owner. One third of that amount – one third of the cost of the fuel saved – went to Boulton & Watt as their royalty. This charge emphasised the status of the Watt engine as a successor to, and an improvement on, the Newcomen engine. Although the calculation seemed simple, it led to no end of wrangles. How exactly should one estimate the cost of the fuel saved? And what manager would not estimate low so as to pay Boulton & Watt little?

Another significant problem with this seemingly simple arrangement was that the mine-owners rarely paid Boulton & Watt for the engines outright. Although Boulton yearned to create a foundry to make complete engines, Watt was at first so jaded with his dealing with navvies and workmen that he wanted nothing of it. There were various alternatives. Occasionally, Boulton & Watt modified existing Newcomen engines, re-using materials and parts. Alternatively, at the beginning, they drew on limited capital, bought in parts, and assembled Watt engines from scratch, in a bid to convince customers of their superiority. Later, owners purchased and put together a kit of parts, designed by Boulton & Watt, but made by recommended suppliers. Watt knew exactly how to work this low-risk scheme: it was exactly what he had done when he was erecting Newcomen engines for Scottish clients. But there was no profit for Boulton and Watt to be had from construction alone, except in one respect: they made the all-important precision valves which regulated the flow of steam.

In fact, Boulton and Watt could make complete engines at Soho only after they created their foundry there in 1796, and by that time they were heading for retirement. That was a good twenty years after they began to colonise the Cornish mines. Until then, most of the company's income, such as it was, came not from the making and selling of engines at a profit, but from patent royalties. Every new engine erected in Cornwall, the 'county of fire engines', returned a steady trickle of money. If mine-owners paid up, and they often did not, that trickle would eventually become a torrent. Initially, royalties flowed only gradually into the company's coffers, while capital flooded out.

Only Boulton's long-term commitment and borrowing power made Watt's idea a practical reality. Time and time again, Boulton was Watt's rock. Until 1775, Small kept the Scotsman flying, consoling him with a stream of encouraging letters as worries about his health drifted to despair of steam and its sickly mechanical progeny. After Small's death in 1775, it was the energetic and long-suffering Boulton who managed Watt's moods, his trips to the Slough of Despond, and his lack of financial acumen. 'Pray morning and evening', Boulton implored Watt in April 1781, 'for you want nothing but a good opinion and confidence in yourself and good health'. Even in 1781, there was much to do before the engine could conquer its new markets: the factories.

· CHAPTER VI ·

DOUBLING, ROTATING, EXPANDING AND INDICATING

By 1778, Watt at last had an engine to his satisfaction. The simple engine conceived in 1769 had been left far behind, and he now had a machine suited to the Cornish markets. But the work was by no means over. In an astonishing binge of creativity, resulting in five new patents between 1780 and 1785, Watt struggled to 'improve' the engine still further. A simple pumping engine that could pull but not push was no longer enough. He wanted a machine that could do both. An engine that delivered a back and forth motion was fine for the mines but barely worth a look for the factory-owners. In 1781, Boulton knew that British industrialists were 'steam mill mad': they wanted a machine that delivered rotary power direct, and Watt must give it to them.

There was more. To save wasted steam, Watt had conceived of a perfect engine relying on a single cylinder of steam per stroke, rather than the Newcomen engine's three or four. Now he, and rival engineers, considered how an engine might make do

Illustration 9. An engine from 1787–8 for the Albion Mill, drawn for Watt's friend and advocate John Robison. This design shows a machine of glorious complexity. The engine is 'double-acting': steam can be admitted to the cylinder (left) both above and below the piston; and it has Watt's 'parallel motion' attached to the beam (top left). More than a pump, this engine is 'rotative' and can drive factory machinery: to avoid the 'crank' of Watt's competitors, it has the 'sun-and-planet' gearing (bottom right). The engine regulates itself: there is a centrifugal 'governor', with its characteristic suspended and rotating balls (above and to the right of the cylinder). (Source: *Encyclopaedia Britannica* (3rd edn, 1797), vol. xvii, plate v, illustrating John Robison's articles 'Steam' and 'Steam Engine'.)

with less: instead of allowing a full cylinder of steam to act as a source of constant pressure, half a cylinder, say, might be left to expand gradually. They might generate more power, with smaller engines, by working steam at far higher pressures. In Watt's attempts to give the engines ever greater 'self-control', he gave them 'governors'; to market them in the factories, he rated them in 'horsepower'; and to make them monitor their own performance, his assistant Southern gave them 'indicators'. Small wonder that Wordsworth gaped at these engines, convinced that they had 'life and volition', and that Coleridge called this mechanism made intelligent 'a giant with one idea'.

Going Round in Circles: the Hunt for a Rotative Engine

In 1766, Watt dreamed of an engine that used steam to generate rotary (circular) power directly – not just a pump to supply a waterwheel whirling on its axis. In the patent of 1769, he gave a teasing description of just such a 'rotative engine'. His 'steam wheel' was the talk of Kinneil. By 1772, Roebuck had greater hopes for it than the humdrum 'vertical engine' with its intractable cylinder. Watt was testing prototypes in Birmingham in 1774, much to the excitement of the excitable James Hutton, who quizzed and chivvied Watt: 'Is your egg hatched yet, or are you still brooding like a bubly jock?' By August, with Boulton's backing, Watt had a six-foot revolving chamber in

Soho, amazing all and sundry but with the annoying habit, surely remediable with a few years' training, of leaking steam. Hutton was convinced that the steam wheel would bring Watt the fame he so richly deserved with the novelty-hungry multitude, and in 1775 the inquiry into the first patent looked carefully, and we may assume admiringly, at this 'curious wheel'. But this project, like the alkali scheme and the surface condenser, foundered. It never worked.

By spring 1775, of course, thanks largely to Wilkinson, Boulton & Watt had bigger and better fish to fry. The Watt engine had surpassed expectations in its ability to deliver reciprocating (up and down or back and forth) power economically. Now there was a possibility that that could be transformed *into* rotary power. Although it was easy to get from reciprocating to rotary *motion*, many times engineers failed in their attempts to extract rotary *power* from the (reciprocating) Newcomen engine. If Watt failed too, his steam engine would be good for pumps and little else. But by the 1780s, consumers of rotary power in the mills, factories and workshops of the new industrial cities were laying off hands, re-tooling with heavy machinery, and screaming for steam.

To understand how Watt responded, we need to remind ourselves how the early steam engines worked. There was a working cylinder, capped both top and bottom to keep in the steam. The piston-rod passed through a 'stuffing box' made air- and steam-tight. Pipes, regulated by valves, channelled steam from the boiler to the space above the piston. When the piston moved downwards, in its 'power stroke', steam pressure pushed it to the bottom of the cylinder. Meanwhile, a valve connected the space under the piston to the separate condenser, which served its purpose in creating the (partial) vacuum. So that the piston could rise up to the top again, after doing its work, the connection to the condenser closed and the so-called 'equilibrium valve' opened. This linked the spaces of the cylinder above and below the piston so that they were brought into 'equilibrium' (the pressure on both sides was the same). Now, it was the weighting of the engine (a counter-weight), not the power of steam, that drew the piston back up to the top, and the whole cycle repeated.

An engine like this was 'single-acting': there was a 'single' power-stroke and then a 'dead time' during which the engine re-set itself, ready to begin again. Like the single-acting Newcomen engine, this machine was best suited to raising water. It could supply a waterwheel and generate rotary power that way. That was what Boulton had asked for when he began to collaborate with Watt and, as Hutton reminded him, he had it by 1775. But all this paraphernalia wasted money, space, and power. There were losses through friction, through transmission from one engine to another, and in the waterwheel itself. Smeaton could find no waterwheel so 'perfect' that it could keep itself in perpetual motion. Better, then, not to wait eternally for Watt's 'steam wheel', but to get rotary motion direct from the best machine then available: the 'single-acting' reciprocating engine.

Suns, Planets and Cranks: Circling Around Pickard

This turned out to be inadvisable, although not strictly impossible, with the simple engine I have just described. The problem was that 'dead time' when the engine was re-setting itself and not developing power. Even sophisticated single-acting machines had an alarming tendency suddenly to turn backwards – with ruinous effects on the new and sensitive spinning machinery. In 1775, after years of expenditure with no return, clinching the extension to the 1769 patent had opened up potentially lucrative markets, especially in Cornwall. Exploiting those markets took work in design, development, marketing and 'after care', and this left Watt little time for further mental gyrations. For a few years, he put the rotative steam engine on the back burner.

There was, however, another possibility. If a single 'single-acting' engine was little use, perhaps two would do the trick. One possibility was to couple two single-acting engines together and have them working 'out of synch' in order to deliver something closer to uninterrupted power. That might then be transformed into the close-to-constant rotary power that mill owners wanted. Watt tried this out in 1779. But he still needed a simple, workable, mechanical arrangement to transform the reciprocating motion of the top of the piston-rod into rotary motion.

A standard solution, taken from the basic alphabet of the mechanic, was to use the 'crank'. In the pilot of 1779, Watt gave each of two engines its own crank to generate rotational power and deployed a special mechanical arrangement to carry each crank past 'dead centre'. According to one account, at least, Watt was on the point of patenting the crank, as applied to the steam engine for this particular purpose. As usual, he built up an 'experimental' model in preparation, to sort out practical problems invisible on the drawing board. No surprise, then, that these materialising thoughts fell under the hungry gaze of an employee of Boulton & Watt who promptly passed the idea on to another engineer grappling with the same problem.

That same year, James Pickard employed the Bristol engineer Matthew Wasborough to construct a Newcomen engine producing rotary motion directly, at the Snow Hill mill close by Boulton & Watt in Birmingham. Pickard had to make do with a Newcomen engine because the partners refused to grant him a licence under their steam-engine patent. Wasborough was wrestling in parallel with Watt, trying all manner of mechanical arrangements, when he decided that one of them (the 'ratchet and pawl') had potential. He patented it in 1779, but there were practical problems and he swapped it, in 1780, for the crank (and weighted gear) which Watt had been using at Soho. Although the crank was common property, applying it to the steam engine to get rotary motion was (almost) new, and in August 1780 Pickard patented it.

This was a crafty move – and a potentially devastating one for Watt. Rotary motion was the new holy grail for the steam engineers. After fifteen years' work, it looked like Watt had been pipped at the post. He was driven to incoherence when he tried to describe just how Pickard managed to get away with it. Watt's recollections were inconsistent, and like many public figures he was not above portraying his actions in the best light. First he claimed that Pickard had stolen his idea. Then he dismissed Pickard's patent outright, claiming that the 'real inventor of the rotative motion was the man, be he Chinese, Indian, Arabian, Greek, or Goth, who first made a common foot-lathe. The applying it to the engine was merely taking a knife to cut cheese, which had been made to cut bread.' The patent was valid but misappropriated. The patent was invalid since trivial.

Either way, Watt was edgy, and pondered how to safeguard the future of Boulton & Watt. He could use the crank – and pay a royalty to Pickard. Boulton & Watt negotiated with Pickard but quickly found the prospect too much to stomach. Watt was too proud, or irritated, and certainly too canny to buy back what he (sometimes) considered to be his own invention. The other option was to capitalise on his vast experience as craftsman, workshop manager and engineer to find new ways of getting rotary motion. Watt was in his element, moving far beyond the basic alphabet of mechanisms and conjuring up a veritable 'theatre of machines' to upstage Pickard. Within the year (1781), Watt had hit upon and patented no fewer than five new methods to get 'continued rotative motion' to drive 'wheels of mills or other machines'. Four never came into commercial use. The fifth was his now-famous 'sun-and-planet' motion, and it stole the show.

The idea was to have two toothed wheels of the same size. One (the sun) was fixed to the end of an axis carrying a flywheel. It could rotate about this axis. The other (the planet) was attached to one end of a rod, the other end of which was connected to the wooden beam, which made its see-saw motion. This second wheel could not turn on its own axis, but was constrained to stay in contact with the first and cycle around it. With the engine at work, the beam rocked, the higher end of the connecting rod moved with it upwards then downwards, and the planet at its other end moved in a circle around the rotating sun. Since planet and sun had interlocking teeth, the sun made two complete revolutions about its axis for every transit of the planet around it. Watt had his rotary motion, and he had invented his way around Pickard and his blocking patent.

Two For the Price of One: Making a Double-acting Engine

There remained the problem of the 'dead space'. Watt was dissatisfied with the arrangement whereby he coupled two 'single-acting' engines. A more elegant solution, of the kind that Watt (harking back to his days as an instrument-maker) found so attractive, was to create a 'double-acting' machine: that is, one in which both the 'down' *and* 'up' strokes developed power. Watt claimed that he began to think about double-acting machines, as a route to rotary motion, in 1774; although, as we will see, he did have a tendency to pre-date his inventions when other engineers appeared with similar ideas. In any case, when Boulton and Watt lobbied to extend their patent in 1775, they presented a drawing of a double-acting engine.

The principle, at least, was simple. To make an engine double-acting, Watt needed steam to be let into the working cylinder first *above* the piston (as usual) and then *below* it. He also needed to arrange for the used steam, above and then below the piston, to rush into the separate condenser. He achieved these aims with a more complex arrangement of valves and pipes. With steam pushing the piston downwards and then upwards, in both cases against a vacuum created by the separate condenser, the engine had two power strokes, one in each direction (and counterweights were no longer necessary).

The benefits were considerable. The double-acting engine which Watt patented in 1782 cost little more than its single-acting predecessor, and it might generate twice the power of a single-acting engine with the same sized working cylinder. Although it did not economise significantly on steam, there were real savings in space and construction costs. Furthermore, by approaching more closely to the continuous delivery of power, the double-acting engine, equipped with the sun-and-planet motion, would make a 'rotative' engine far more likely. But one major obstacle remained.

Elegant Geometry and Ingenious Mechanism: Watt's 'Parallel Motion'

Pulling was easy, but Watt needed to push as well. In the early single-acting engines, the top of the piston-rod and the rocking beam were connected using a heavy chain which wrapped around a 'sector' (shaped like a slice of a circular pie) attached to the beam. This worked because the piston-rod and chain were always under tension. Either the piston moved downwards (as steam was admitted above it) and pulled the chain which in turn rotated the beam or, its work done, the counterweight pulled the chain and piston up again. The chain pulled, but it never had to push.

With a double-acting engine, there was a problem. Both down and up strokes were power strokes. The piston rod had to pull the beam end down *and* push it up again. A single chain was not enough for this. (You can pull with a piece of string, but try using it to push something along.) Engineers looked for solutions. Perhaps the (already) rigid piston rod could be made longer to tower above the level of the beam.

The first chain was still there, connected to the top of the sector; now add a second one connected to the bottom of the sector. The old chain could pull the beam down and the new one could pull it up. Another idea was to give the sector 'teeth' and the piston rod a 'rack' engaging with these teeth. Neither of these schemes worked well enough for business.

Watt had another idea. He dispensed with the chain, took a rigid link instead, joined its lower end to the top of the standard piston rod, and its upper end to the beam. Here was a way of transmitting power by pulling downwards and pushing upwards. So far so good: but simple geometry indicated that this link was usually deflected from the vertical. Whenever that happened, the top of the piston rod would be pulled and pushed a little from left to right. Vibrations like that led to wear and tear, especially on the sensitive but crucial 'stuffing box', and Watt's customers needed an engine that would keep working, night and day. In 1784, Watt tried to get around this problem, one of practice rather than geometry, by using a guide to keep the piston rod in its proper upright position – but that was a temporary measure.

Far better, in Watt's opinion, was the famous 'parallel motion' dating from the late summer of 1784. This was a combination of rods hinged in such a way that power could be transmitted but the piston rod did not vibrate. Watt's idea was to take two rods, each able to rotate about one fixed point, and connect their closest ends, each moving in a circular arc, by a smaller link. There was a point (A) on the link that traced a curve, part of which was very nearly a straight line. This arrangement was known as the 'three-bar' motion: there were two rods and a link. In a full-scale engine, one of the rods was really the wooden beam rocking on its fixed axis. Connecting the top of the piston rod to the point A on the connecting link solved the problem of transmitting power without undue vibration – since A moved in a straight line – but it introduced another one. To accommodate the additional lower 'bar' meant extending the engine-house considerably. Watt worked around this problem by adding to the 'three-bar' motion a parallelogram (a four-sided figure with opposite sides parallel), one lower corner of which reproduced the vertical straight line motion of A. Connecting the piston rod to that corner saved space and completed the 'parallel motion'.

Watt was inordinately pleased by the way this elegant geometrical trick, made concrete in 'one of the most ingenious simple pieces of mechanism I have contrived', avoided all manner of 'pieces of clumsiness'. Years later, he was still telling his son that he was 'more proud of the parallel motion than of any other mechanical invention I have ever made'. With the parallel motion, he had at last achieved a vital goal. He could transmit the power of a double-acting machine to the beam and, with the sun-and-planet motion, transform it into the rotary power that factory-owners craved.

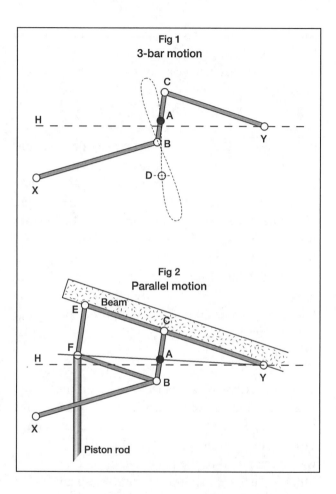

Fig 1
3-bar motion

Fig 2
Parallel motion

Expanding Steam: (Not) Making a High-pressure Expansive Engine

When Watt was working on the Newcomen engine from the winter of 1763–4, he had become obsessed

Illustration 10.
Figure 1. The three-bar motion. XB and YC are bars free to rotate about X and Y respectively. BC is a link joining the two bars. The point A on the link BC moves on the curve shown. Remarkably, between A and D this curve is very nearly a straight line. If HY is horizontal, A will move vertically with very little deviation.

Figure 2. The parallel motion. Here, the bar YC is extended to YE. A second link FE, the length of BC, and a bar FB, the length of EC, complete the parallelogram BCEF. The length EC is chosen so that F is on the extension of the line YA. With this arrangement, known as a pantograph, the point F copies and 'scales up' the motion of A, moving almost vertically. By making F the top of the piston rod and YE the lower surface of the beam, Watt could make his engine double-acting.

(Source: Based on diagrams in H.W. Dickinson, *James Watt: craftsman and engineer* (Cambridge University Press, 1936; new edn, David and Charles, 1967), pp. 138–9.)

---

with creating a perfect engine which wasted no steam. He tended to think of the steam almost as if it were a cylinder's worth of pressure pushing, with constant force, against the piston. But there were other ways of using steam which might lose the advantage of constant pressure and yet, paradoxically, save even more steam. Would that create a hyper-perfect engine?

By 1769, and perhaps even by 1767, Watt began to think about lessening the consumption of steam, and fuel, by using the steam 'expansively'. Originally, steam was let into the working cylinder for the entire stroke. The steam pushed, at nearly constant pressure, as the piston moved downwards to accommodate it. With 'expansive action', however, a limited quantity of steam was allowed into the working cylinder for just a fraction of the stroke. After that, 'cutting off' the supply meant that no more steam entered. The stroke continued and the fixed quantity of steam expanded freely against the piston, reducing in pressure as its volume increased. Watt tried out expansive action with an experimental model in about 1777, and patented the technique in 1782. The first date is disputed and the patent was an odd move – as we will see later.

A simple example (ignoring the insights which would come, later, from the science of thermodynamics) indicates why this was an idea worth trying. Imagine a cylinder with a cross-section of one square foot (or 144 square inches). Say that the stroke of the piston (here signifying the total distance it moves inside the cylinder) is two feet. Suppose that steam is let into the cylinder at a pressure of two 'atmospheres', or very roughly 30 pounds per square inch. In this ideal example, suppose that the condenser produces a perfect vacuum exerting zero pressure. The 'load' on the piston is its area multiplied by the pressure on it, or 144 multiplied by 30, which is 4,320 pounds. Moving through the stroke of two feet gives $2 \times 4,320$ or 8,640 foot-pounds of work, for a consumption of (2×1) cubic feet of steam at a constant pressure of two atmospheres. That's the situation without using the steam expansively.

Suppose, now, that instead of allowing the steam to escape into the condenser after the piston has moved two feet, the cylinder is extended by another two feet. No more steam is admitted. But the steam originally allowed in continues to press down on the piston, with the same perfect vacuum on the other side. The pressure pushing the piston gradually reduces during the final two feet of the stroke. At the end of the stroke, the steam has doubled its volume and, therefore, halved its pressure to one atmosphere. The mean pressure in the second two feet works out at 20.8 pounds per square inch. So, $144 \times 20.8 \times 2$ (or 5,990) additional foot-pounds of work is obtained with absolutely no additional 'expenditure' of steam. Allowing steam to expand to twice its original volume in this hypothetical case leads to $8,640 + 5,990$ or 14,630 foot-pounds of work. That's the situation using steam expansively, and it means, in this example, 69% more work.

If steam was conceived of primarily as a source of pressure, which for Watt and his contemporaries it was, using it expansively had clear advantages – assuming, of course, that an enlarged cylinder was acceptable, and that the smaller, and perhaps slower, delivery of power in the 'expansive' phase was too. But more arithmetic suggests that using steam expansively worked best with pressures many times that of the atmosphere. Experience confirmed this. At the time of Watt's death, condensing engines used steam at only five pounds per square inch above atmospheric pressure, and consumed between five and eight pounds of coal per horsepower per hour. By the end of the nineteenth century, hyper-efficient engines for ships, designed to minimise the weight of fuel carried, approached pressures of 250 pounds per square inch in order to reduce consumption to a paltry 1¾ pounds per horsepower per hour.

Although Watt, like most of us, could not see into the future, we still need to ask why, unlike his more radical contemporaries, he favoured low pressures, when he had been obsessed with saving wasted steam and fuel. There were several reasons. Watt inherited a culture of low-pressure steam. Steam at low pressures was familiar to those used to Newcomen-type engines, like Watt, who had erected Newcomen engines in the 1760s. Low-pressure steam was easy to come by, routine to manage, and safe to use. Increasingly, Watt became concerned with the power of the engine, rather than simply its fuel economies, and that depended on the *difference* between the pressure of the steam and the pressure of the (partial) vacuum created in the separate condenser. With a very 'good' vacuum to push against, his engines could and did use steam at pressures *lower* than the atmosphere. With high-pressure steam, the goodness of the vacuum became less important. But it was the separate condenser, with its perfect vacuum, that helped to define Watt's perfect engine and marked his giant leap towards it. He was personally committed to thinking of the steam engine in that way. Without that commitment, his engine might well not have been completed.

In any case, high-pressure steam was novel, it burst boilers and it threatened life (or so its detractors claimed). Ever since his syringe experiments in the early 1760s, Watt had been nervous of high-pressure steam. There were technical reasons to avoid it. Making the joints, valves and stuffing box steam-tight was hard enough at low pressures. With materials and skills already pushed to the limit in a complex – even over-complex – engine, Watt considered the extra challenges of high pressure at best expensive and at worst utterly impractical. In any case, one major explosion, at unknown human cost, might completely undermine confidence in Boulton & Watt, jeopardising the future security of the company.

Nevertheless, Donald Cardwell calls this Watt's Achilles' heel. It was reasoning like this that gradually transformed Boulton and Watt from decisive innovators to technological conservatives, opposing innovations even among their own staff if they could not see investments rapidly repaid. In November 1785, a pessimistic Watt lamented: 'I find it now full time to cease attempting to invent new things, or to attempt anything which is attended with any risk of not succeeding … Let us go on executing the things we understand and leave the rest to younger men, who have neither money nor character to lose.'

An Engine with Self-Control: the Centrifugal Governor

It was not yet quite time for Watt to pull on his slippers and gaze contentedly into the fire. Boulton & Watt had conjured into life a tremendous machine that moved and breathed like a living giant. But that mechanical beast had yet to be taught to control itself. An engine like Newcomen's, that still needed regular attention from human eyes and ears, was anything but perfect. Watt was adamant that even the operations of his little experiment with the digester in 1761–2 could be 'done by the machine itself'. Although much of the complex cycle of operations in the mature Watt engine had been automated, Watt recognised that more could be done when he introduced a device with the wonderful name of 'governor'. An engine with a governor had more than a modicum of self-control.

Governors had already been used in flour mills, powered by wind or water, to adjust the distance between 'nether' and 'upper' millstone as the working rate of the mill changed. Without a device of this kind, a vigilant human minder was required to regulate the distance between millstones manually. Skilled workers fell sick, putting mills out of action, sometimes made mistakes, or demanded high wages. Machine regulators may not have been perfect, but they were cheap, consistent and uncomplaining. A common solution was the 'centrifugal governor', so-named because it relied upon two spinning balls, rotating about an axis, and raised or lowered according to their speed thanks to a varying centrifugal force (one directed away from the centre).

Some inventions popularly attributed to Watt were not his sole creations. The governor was one of them. Watt exploited a reservoir of knowledge dispersed among the machines and makers of existing power technologies. A 'steam governor', similar to those used in flour-mills, would serve for the steam engine. If the engine functioned at too great a rate, centrifugal force drove those rotating balls outwards and (usually) upwards, which in turn operated a steam valve that slowed the engine. In this 'feedback' mechanism (as we would now call it), information transmitted mechanically from the engine to the governor fed back to regulate the engine. Once again, unruly machine-minders were cut out of the mechanical loop. This action, in which the engine appeared to be self-governing, stunned contemporary commentators, who likened the machine to an intelligent being.

Although it is unclear quite when Watt first used a 'governor' with the steam engine, the earliest histories claim that the efficient rotative engines that Boulton & Watt were making from 1784, including one for Whitbread's brewery, had all 'mod cons': the sun-and-planet motion to avoid Pickard's crank; the parallel motion essential to make the engine double-acting; *and* the governor for its self-regulation.

Although Watt was 'governing' the steam engine before Thomas Mead patented his own 'regulator for wind and other mills' in 1787, he was really only re-using a principle that had been common in mills for some time. Perhaps he was 'taking a knife to cut cheese, which had been made to cut bread'. He neither patented the governor nor claimed to be its inventor. What he did do was recycle a practical idea and fuse it to the steam-engine with astonishing effect.

Standardising Steam: Taking the Measure of 'Horsepower'

While the governor gave the engine self-control, men of science and practice alike looked to standards of measure to rationalise and control their own chaotic environments. Watt's Lunar Society friends engaged in the Birmingham Assay Office wanted accurate, physically invariable, and universal standards of measurement when pounds and ounces swelled or shrank from one city to the next. In November 1783, Watt grappled with experimental scientific results tabulated according to different national measures, and asked how they might be made to 'speak the same language'. Better, he thought, to agitate in Britain and France for a 'philosophical pound', underpinning a set of internationally valid units of weight and length, with decimal divisions for easy calculation. In the 1790s, the French would impose the now familiar 'rational system' of *mètre* and *kilogramme*. But for years British engineers, too, had been crying out for standard measures of steam power, easy to translate from place to place.

Managers avid for economy needed information. When, in the years up to 1780, an atmospheric fire engine or a steam engine simply pumped water out of a mine, or raised water for a waterwheel, or blasted air for a furnace, there were routines for measuring the work it could do and then comparing it with other engines. Smeaton assessed a Newcomen engine by measuring the volume of water it could raise in a minute and multiplying it by the height raised in feet. It was only a short step to measure the weight (of water) in pounds raised to a certain height. This quantity in 'foot-pounds' was known first as the 'effect' and later, technically, as 'work'. Dividing by time gave the engine's rate of doing work, familiar to us as its 'power'. On the other hand, the 'duty' (Watt's term) of a fire or steam engine was a measure of how much weight could be raised to a height by burning a bushel of coal. This measure imparted valuable information to mine- and mill-owners, since it linked work to cost.

As soon as the reciprocating Watt engine was estab-lished in the market for lifting and pumping, Watt needed to convince a new cohort of cost-conscious clients, managing mills and factories but ignorant of the steam-pumping culture of the mines, that a rotative steam engine would best satiate their lust for power. Using a rotative engine to drive machinery directly instead of pumping water to drive a water-wheel could cut costs dramatically, and the managers soon cottoned on. But the exact 'measure' of the 'goodness' of an engine had direct repercussions for purchase cost, royalties (or premiums) paid, fuel consumption, and the all-important anticipated power output. Finding a comparative measure of the 'goodness' of engines for factories was fraught with difficulties. Measures good for the mines did not work for mills. Coupled to complex machinery for the preparation and spinning of yarn, a rotative steam engine was doing more than lifting a weight to a height in a given time. Each machine had a different 'load', to which the engine's power should be ideally matched. An under-powered engine left machinery idle. An over-powered engine under-used was a waste of money and, at this crucial early stage, Boulton and Watt had no wish to 'heap expenses on an infant manufactory'.

In this context, a traditional source of power, and the competitor for the engine-builders, was the beast of burden. Comprehending, and marketing, the new technologies meant comparing them to ancient ones like the horse itself. It made sense, then, for Watt to follow Thomas Savery and other technical authors, and to make the 'horse' the standard by which to measure the effectiveness of the steam engine. Admittedly, horses got tired and ill, came in all shapes and sizes, and had been bred up for different attributes like speed, or strength, or stamina. If there was no standard horse, there might be an average horse. Savery decided that to have one horse available for work all the time, it was necessary to have three at hand. He was less interested in finding the amount of work a horse could do than in the fact that if his 'miner's friend' could do the work of four horses driving a pump, say, then the engine was a '12-horse engine'. This comparison of horse and engine smudged the distinction between animal and machine, even as it showed how to compare them.

Ironically, it was the inconstant horse that would be the common measure of machines. Watt claimed that an 'average horse' could work for several hours raising 22,000 pounds one foot in one minute. Coining a new term, he fixed one 'horsepower' at 33,000 foot-pounds per minute – that is, adding half again so that no one could complain that an engine rated at, say, 'five horsepower' was giving less than its proper 'horse value'. For a rotative Watt engine, the 'nominal horsepower' was then established from the dimensions of the engine cylinder, the length of the piston stroke, and an empirical average of working pressure. In 1785, managers outside London paid a premium of £5 per 'horsepower' per annum (typically £200 a year) for engines covered by the 1769 patent. Establishing the 'horsepower' as a common unit of measurement among prime-movers solved, in principle, the economic problem of comparing Watt's engines, and especially his rotative ones, with all *possible* engines, whatever their construction. By taking this universal, rather than local and practical, view of machines, Watt made himself a 'philosopher of mechanism'.

A Self-registering Engine: Southern's 'Steam Indicator'

If the governor gave the steam engine self-control, and 'horsepower' rated it in ways that factory owners wanted, the 'steam indicator' made it self-registering. With the indicator, owners could measure engine power reliably and correlate that power with the exact demands of factory machinery.

From the early 1790s, Boulton and Watt started to use pressure gauges or 'indicators', based on mercury barometers with a pointer moving along a scale, to measure the varying pressure inside a steam engine's working cylinder. Although the main concern was the maximum and minimum pressures (from which a guess at power could be made), even these were hard to register by eye with a rapidly working engine. There was real interest in these devices from Watt's friend, the cotton-spinner George Lee, and from the Leeds mill-owner Benjamin Gott. Both had rotative Watt engines by 1792 but, Lee complained, 'a physi^n can give no good account of ye state of his Patient's health without feel^g his Pulse & I have no Indicator'! Rather than guess what size of engine was required – a common estimate was one horsepower for 100 spindles – Gott and Lee used the indicator to monitor the effective engine pressure needed to drive different kinds and quantities of textile machinery: the successors to Arkwright's spinning frame (originally horse-driven); carding machinery patented in 1775; and roving machines replacing manual labour.

Of course, the pressure difference acting on the piston was not constant throughout the cycle. In April 1796, John Southern, by then Watt's co-partner, worked out a way of recording the variation continuously. He replaced the pressure-gauge pointer with a pencil which traced a line on a paper card moving in line with the piston. The resulting graph, known as the 'indicator diagram', mapped the cylinder's pressure against the piston's displacement. Because the cycle of the engine repeated almost exactly, the graph formed a closed loop (like a circle, or even a figure of eight). From the loop's area it was at last possible to find out the work being done. Engineers could use the diagram to optimise the engine's performance. Varying the timing of the valves, for example, could seriously enhance the engine's power – as revealed by the indicator diagram. Boulton and Watt's friends made the most of this powerful technique, created by Watt's talented assistant. Lee admitted that he was like 'a Woman dying to hear (or tell) a Secret – to know Southern's Mode of determining Power'.

In fact, the indicator was just another in a series of automatic, self-acting devices: machines like Watt's 'locked-up automatic counter' used to record the number of strokes made by an engine over a long period; or his French *ami* G.R. de Prony's 'dynamometer' which automatically measured the output of power from all kinds of engines; or devices for registering intervals of time too tiny for humans to judge. But as a machine for making diagrams, the indicator was also part of a growing trend to 'picture' numerical data extracted from the chaos of science, economics, history – and the practical arts. Better than wordy descriptions or subjective claims, machine-made graphs of quantitative data appeared direct, unambiguous and objective. Collecting and disclosing information independently of wilful human observers, such devices were as reliably honest and informative as only machines could be.

Taking stock, it's clear that there were many technical and administrative problems to solve before Watt's double-acting rotative engine 'worked'. But there was so much more to 'working' than the purely technical. To be sure, Watt perceived a set of technical problems and, with the support staff at Soho, he solved them. Some innovations served to make a practical engine more efficient than Newcomen's, or even Watt's 'single-acting' engine, and one that could do more than simply pump water. Other inventions dispatched rival engineers. Self-acting devices, especially, put skilled operators out of a job, saved owners' money, or convinced customers that this was the prime-mover for them.

More than this, in the 1780s and 1790s, Boulton & Watt engineered a complex network of mechanical contrivances (and, equally, techniques for marketing and maintenance) to be produced and sustained by skilled personnel. Any weak links jeopardised the entire project. To understand this is to understand that there was nothing inevitable about the commercial success of the steam engine with separate condenser. And perhaps the biggest threats to Boulton & Watt's success were the pirates.

CHAPTER VII

CIRCUMNAVIGATING WATT:
PIRATES AND PATENTS

Leaking trade secrets was a major problem in the late eighteenth and early nineteenth centuries. Stealing them was a major sport. A veil of secrecy surrounded Keir's lucrative new process for synthesising alkali, bubbling away at the Tipton works. In 1765 Watt chided Robison for careless talk, in 1768 Boulton was sworn to silence by Roebuck, the Kinneil experiments were clandestine, the Lunar circle met secretly, and in 1779 Watt was convinced that the steam-engine crank had been passed on to Pickard by an eagle-eyed chap haunting Soho. By 1785 Wedgwood and Boulton had had enough, and they led a 'Chamber of Manufacturers' formed to fox the spies, especially the foreign ones, snooping around their factories – and even sizing up Watt's adolescent engine. Boulton and Watt told friends, never rivals, about the indicator diagram. In the new century, one suspected Bonapartist spy, de Longastre, pandered to the vanity of Keir and Watt, pastelling their portraits while he waited, perhaps a little late in the day, for give-aways in industrial chit-chat.

The Lunar Society members were not the only industrialists who walked a tightrope, ruthlessly uprooting technological espionage even as they showcased their modern factories in Soho, Etruria and Tipton – and quietly charted the progress of their rivals. Watt himself had explored Boulton's Soho works before moving to Birmingham, and there was a delicate balance between sanctioned swooning at safe selected highlights of a glistening factory show-space, and trespassing on the private parts of industry, smeared with the dirt and danger of evolving designs. Big experiments in industry were hard to hide. James Watt Jr summed it up later when he called open exhibitions of new processes 'suicidal folly' – and demanded a resumption of 'our ancient habits of privacy and exclusion'.

One solution was to use patents as a legal basis for the prosecution of industrial pirates, even though many engineers, among them Isambard Kingdom Brunel, decried patents as stifling competition. Watt's knowledge of patents would eventually become second to none. In 1769, 1775 and again in the early 1780s, he had diligently obtained patents in Great Britain for the many mechanical inventions that made his steam engine versatile and bankable (and some that he was convinced would never work, but wanted to block others from pursuing). He secured patents or manufacturing 'privileges' in France, Spain and the Netherlands, and he applied for similar protection in America, Prussia, Belgium and Austria.

In 1785, he appeared as an expert witness to support Richard Arkwright, inventor of the 'water frame' that had mechanised spinning. After witnessing Arkwright's patent being thrown out, watching messy legal battles over Keir's alkali-making, and experiencing the horror of gruelling encounters over his own patents in the 1790s, Watt pushed to reform British patent law which, by then, was in nightmarish disarray. He even lobbied the Lord Chancellor. Some of his ideas made it into law, even if a complete overhaul had to wait until 1852.

By then it was too late to help him – or his opponents. Although Boulton & Watt's licensing terms were considered fair, some mine-owners and manufacturers made themselves independent of the company by exploiting cheap and simple, if relatively inefficient, Newcomen engines of improved designs. But the rewards of steam were high. A healthy fraction of those keen to economise on fuel simply paid up for the Watt engines. Some manufacturers and mine-owners ignored the patents, obtained copies of the engines from other suppliers and paid no royalties (or premiums) to Boulton & Watt (who never granted licences for other engineers to manufacture Watt engines in Britain). Other canny mechanics challenged Watt's leaky patent of 1769, perhaps even intending to puncture it fatally by promoting new designs which relied on the separate condenser but which dared to be more efficient than Watt's engine. Soon there was a master list at Soho of these patent pirates.

Watt was painfully aware that the very foundation of his growing business empire, the 1769 patent, was the quirky offspring of a talented mathematical practitioner rather than the pedantic child of a nit-picking lawyer. Worried about Arkwright's fate, he told Boulton: 'all the bells in Cornwall will be rung at our overthrow'. Anonymously and nervously, the partners began to prepare their own defence, while keeping a close eye on new steam patents. Not paying up, or infringing Watt's patents by re-using the separate condenser in new ways, stirred Watt to anger. When Boulton & Watt could not harass miscreants into submission, they considered legal action. Boulton and Watt knew that court cases, with their special juries, witnesses and experts, whether eminent or discreditable, were extremely disruptive affairs. Where other pressure did not work, they were necessary to extract withheld premiums or to block opponents, but they also turned the spotlight on to their monopoly. A decision on that point against Boulton & Watt would have meant the weakening, if not the demise, of the firm.

Consider the case of Jonathan Hornblower Jr (1753–1815), plumber turned engineering revolutionary. As one of a family of engineers, Hornblower was far more of an insider than Watt had been. Hornblower had his Boulton in John Winwood (1732–1810). In 1781, the engineer patented a pumping engine. It was double-acting, a full year before Watt's patent of 1782. Even more intriguingly, it had not one but two working cylinders linked in sequence, an idea conceived in 1776. In what came to be known as a 'compound engine', steam entered the first working cylinder as usual and then, instead of doing its work and heading off to the separate condenser, it passed to a larger cylinder where it expanded against the piston to develop more power. Expansive working made better use of steam and led to healthier bank balances. With hindsight, this was the future of steam. With high pressure, expansive action and 'compounding', Arthur Woolf (1766–1837) would achieve fuel economies far surpassing those of Boulton and Watt. It was these engines that came to dominate the red-in-tooth-and-claw Cornish markets, French industry in which coal was scarce, and, later still, marine engineering: 'triple expansion' engines for ships issuing from Clydeside yards would notch up a third cylinder and even better efficiencies.

Back in the 1780s, Watt had for once – or should we say again? – dragged his feet. Only in 1782 had he thought to patent the expansive use of steam, a year after Hornblower, and he made little use of this principle in practice. Boulton and Watt were more concerned to send out friends and employees to investigate Hornblower's engines, which constituted a serious challenge to their near monopoly in Cornwall. John Southern thought the first such engine, erected in 1782 at Radstock Colliery, 'wretched'. But Watt wanted to open all eyes to such (alleged) 'infringements made upon his property'. He trod carefully with the Radstock owners, musing that 'one should not warn a man that we mean to break his head, lest he put on a helmet'. Then he tried to scare off others from straying to the Hornblower camp with an ad in the Bristol papers stating – or rather, overstating – the several inventions marked out by his own patents: the air-pump was in common use in atmospheric engines; expansive action and the double-acting engine were both covered in Hornblower's earlier patent. In private, Watt understood that Hornblower's engine was likely to be more efficient, and moaned: 'If they are protected in it we may give up engine making.' He had reason to worry, since in those years engine sales were minimal: just one in 1781 and two in 1782. Watt took solace in the fact that these 'Imps of Satan', as he liked to call them, would be 'duely rewarded ... by their patron Belzebub'.

In 1792, anxious to recoup development costs, Hornblower copied Savery, and Boulton and Watt, and applied for an Act to extend his patent. Despite many contrary accounts, that patent had never been legally challenged. Nor was Hornblower tried in court and proved a pirate. On the face of it, Hornblower's case for an extension was just as strong as Watt's had been. Cornish engineers had no doubt that the new engine was superior, delivering up to 50% more power – and as one of Hornblower's supporters put it, 'the public ought not to lose the advantage', merely because Boulton and Watt were first on the scene.

Boulton sprang into action. His undiminished and devastating skills as a parliamentary lobbyist ensured that Hornblower's campaign came to nought. The application to extend was withdrawn. Watt's gain was Cornwall's loss.

Meanwhile, another Cornish-based engineer, Edward Bull, looked for a way to get rid of the great 'see-saw' beam which transmitted power from piston to pump in the vast constructions of the traditional Newcomen engines and their newer rivals. Starting in 1792, and working with Richard Trevithick, Bull put up no fewer than ten engines, in each of which he inverted the working cylinder over the mineshaft in order to connect the pumps directly to the piston. In fact, Watt had tried the same topsy-turvy trick in a tin-plate model engine back in 1758. Beamless engines like this were indeed simpler, more compact, and cheaper than Watt's – but they still needed the separate condenser to be fuel-efficient, and that was like a red rag to a bull for Watt's lawyers.

In one of the most significant patent actions of the Industrial Revolution, Boulton and Watt went to the Court of Common Pleas in 1793 avid to stamp out infringements or, better still, to make the perpetrators pay for past misdemeanours. Thanks to a special jury packed with friends, collaborators and employees, they won an injunction against Bull, who admitted his fault. However, the alarming decision of the court, on a majority verdict, not to confirm the validity of the amateurish, catch-all specification that was the

1769 patent, unsettled Boulton and Watt mightily, not least because it led to an increase in piracy. Boulton and Watt pulled out all the stops, recruiting scientific friends, employees, fellow businessmen and sympathetic politicians to win the day, in part by painting Watt as a 'philosophical inventor' who was above all pettiness and worldly things. It was a long haul, and a close call. Only in 1799 did the Judges of the King's Bench tardily but decisively find for the patent and for the partners. Hearing that Boulton and Watt had finally pummelled the pirates into submission, Watt's old friend Joseph Black wept with joy.

In the interim, James Watt Jr and Matthew Robinson Boulton had travelled incognito in Cornwall as spies for their fathers, or bare-faced to frighten miscreants into coughing up, allegedly to the tune of £40,000. A network of friends in London and the provinces, including Southern and Murdock, snooped around, peering over walls and wandering through factory premises to check, and check for, piracy. Shockingly, one of the worst offenders was John Wilkinson. When his irate brother William grassed on him after a quarrel, it transpired that John Wilkinson, the man who had played such a pivotal role in making the engine viable in 1775, had slipped from grace, surreptitiously making engines for himself and others, and even encouraging other manufacturers to follow suit. Faced with the evidence, Wilkinson could do little but capitulate. Most worryingly, this case underlined the fact that Boulton and

Watt's policy of releasing crucial diagrams in order to have parts made outside Soho was dangerous, as was the repeated haemorrhaging of expertise through disenchanted and defecting employees.

Jabez Carter Hornblower (1744–1814) – a man often confused with his brother Jonathan – and J.A. Maberley had apparently coaxed crucial facts from a 'drunken scoundrel' who once worked at Soho. Theirs was a tricky case. When confronted as plagiarists, they tried to bargain, begging amnesty in return for a pious promise to infringe no further. Boulton and Watt pondered, anxious to avoid another disruptive suit, but concerned that apathy would encourage and unleash a 'tribe of pirates'. The case went ahead in 1796, again through the Court of Common Pleas, and again with Boulton and Watt's cronies on call. They were more than a match for Hornblower's allies, all of whom Watt Jr claimed he could prove 'ignorant, interested, or malicious'. The partners won, then lost at appeal, then, by the skin of their teeth, won again when the King's Bench upheld the 1769 patent.

Estimates of legal costs incurred to secure their patent and 'well-tried friend' range upwards to a staggering £6,000. They even made models of the Hornblower and Maberley engines to convince the court. But it was worth it. At last Boulton and Watt could safely retire, raking in the vast royalties owed to them, their fortunes well and truly made thanks to very personal vendettas against men like Jabez Hornblower, who was left to rot in a debtors' prison.

The 'perfection of Mr Watt's engine', he seethed, had been used to 'subordinate every other professional man in the scale of comparison, to repress the energies of his contemporaries, and give a deadly blow to competition'.

· CHAPTER VIII ·

MANUFACTURING AND MARKETING: THE BUSINESS OF THE STEAM ENGINE

During the eighteenth century, some 2,500 steam engines were manufactured. Boulton & Watt produced about 30% of those made in Britain between 1775 and 1800, during their years of monopoly. That meant, in round terms, 500 efficient Boulton & Watt engines, of which more than 300 were rotary engines, contributing to that 'steam revolution' in British industry. Since so many of the engine parts, and associated technologies in mines, factories and mills, were constructed from iron, there was a large increase in demand. This was just one of the ramifications, beyond brute power, of the steam engine's success.

These bald economic facts tell us nothing about *how* the engines were disseminated, and little about their consequences. There is another story to be told about how Boulton & Watt went from satisfying close collaborators like Wilkinson to serving 'all the world'. It's easy to forget that the engine business became a serious money-spinner only after 1787, more than a decade after Boulton and Watt had joined forces.

For that to happen, the partners had to take a niche product, develop it as a near-universal source of factory power, and even showcase it themselves in model mills. They had to create and conquer international markets, fraternising with foreigners while resisting the brain-drain that saw friends taking up lucrative contracts abroad. They tardily gathered funds to manufacture complete engines rather than contracting out to other dubious suppliers (like Wilkinson). To ensure the longevity of the firm, they needed to manage their workforce and train up their sons as captains of industry.

Spinning Steam

In the early 1780s, the most exciting market for steam, and potentially the largest, was the mechanised factory. Now that the factories of industrialising Britain are so closely associated in the imagination with steam power, it is hard to grasp what a momentous innovation this was. The 'dark satanic mills' of the early Industrial Revolution were not all so very gloomy. Horsepower came, precisely, from the brute power of the horse. And places like Robert Owen's famous New Lanark in Scotland, combining social experiment with rational production, sat by fast-flowing rivers, taking the advantage of raw natural power for free. Boulton was not the only factory-owner who saw that steam could supplement water power in summer when torrents became trickles, or in winter when water turned to ice. And the rotative engine was more than just a supplement to water power or to the horse gin. It meant that mills could be relocated to the cities, with all that that entailed.

Boulton and Watt moved quickly. Once they had a working rotative engine in about 1783, they began to equip factories using machines, not men, to grapple raw cotton and flax and turn out finished cloth. Boulton and Watt contracted to build their earliest rotative engine for a Nottinghamshire cotton mill in June 1785. As with many new technologies, problems arose during construction: even Watt admitted that he needed four or five months to get an engine up and running. And although the partners expected this venture to generate good press, and more orders, the mill-owners scuppered that hope by refusing to let anyone see the engine.

To publicise this experiment in industry, the partners instead established a thoroughly modern mill. Boulton and Watt opened their own Albion Flour Mill on the Thames, in the very heart of London, in 1786. With two patent rotative engines powering novel machinery designed by John Rennie (1761–1821), it was the talk of the town – and a hugely successful publicity stunt. Here was a 'show-piece' factory rivalling Soho and Etruria, grabbing attention from the great and the good at a splendid opening ceremony and a more considered, but still positive, response from those less fashionable visitors that continued to flood in. What better way to make the case than to go ahead and do it?

Worlds apart from the local mills squatting on village streams, industrial-scale steam-powered mills, fitted out for the new industrial classes, only needed a good supply of coal – although they tended to cluster on large rivers and along the coast in order to get to their customers easily. Steam factories could likewise come to the major centres of population. Populations, too, followed work. Some of the working villages were little utopias – like textile magnate Titus Salt's 'Saltaire' near Bradford, with its educational institute (but no pub) for recreation, and a church for contemplation. Charles Dickens painted a ghastly caricature of soulless industrial worlds in his *Hard Times* (1854).

Inventors were quick to yoke the steam engine to the machinery for spinning cotton and then for weaving as well. Glasgow illustrated this perfectly. By the early nineteenth century, cotton had taken over as the key industry. By the time of Watt's death, Glasgow boasted 2,800 looms in eighteen steam-powered weaving factories concentrated near the city centre. As the cotton industry grew, so did the market for steam, with industry and mechanism developing in tandem. Factory-owners rejoiced in automation: traditional skills and expensive workers gave way to cheap mechanical production.

There were less happy consequences. Goods might be cheaper but handloom weavers, no longer owners of the means of production (their looms), saw themselves reclassified from craftsmen to semi-skilled employees. Reduction in the demand for labour, both skilled and unskilled, might easily lead to a fall in wages as men and women fought over a smaller number of jobs. Communities built upon traditional practices might disintegrate, to be replaced by new ones marching to factory time. The steam revolution did much to usher in mass production in large factories and the demise of independent craftsmanship. While politicians gleefully pointed to the economic benefits, social commentators deplored the loss of individuality and impoverishment of the human spirit.

Global Steam

Boulton wanted to 'make for the whole world'. How, then, could Boulton & Watt export beyond Birmingham, beyond the lucrative but limited Cornish market, outwards to continental Europe, and across oceans to the sugar plantations? In fact, the Europeans were at least as receptive to the new technology as the conservative British. France had a Watt engine as early as 1779. Germany followed suit in 1788. Dutch engineers saw steam as the solution to their national drainage problems and in 1786 the Batavian Society of Experimental Philosophy (founded to promote industry in 1769) paid out for a fifteen-year monopoly on the use of Watt's inventions throughout the Low Countries. Within a year, a Watt engine was pumping water from the Blijdorp Polder. Foreign interest was palpable and, after the new technologies had been proved on home territory, their adoption abroad was little short of astonishing.

Boulton and Watt were prepared to reach out to these attractive markets. One rather comforting image of Watt is the horny-handed son of toil, ignited, as if from above, by a spark of 'mechanical genius' but lacking the most basic cultural accomplishments. Not so. He learnt German and Italian, not to immerse himself in fine words but to gut contemporary classics on mechanics and practical science. He was at least as proficient in French. This linguistic facility, although hard-won, was increasingly common. Mathematical practitioners and 'civil engineers' mastered foreign tongues as a matter of urgency.

Sometimes selling the engine abroad brought tempting offers to sell out. Watt was hotter property than the engine itself. Friends in Scotland tried to keep hold of him, in the face of foreign governments trying to poach him in an eighteenth-century brain-drain. When the engine was still an 'uncertainty', Hutton advised Watt to brag about it and to threaten to leave Scotland unless they made it worth his while, perhaps with the post of 'Ingeneer' under the Board for the Improvement of Manufactures and Fisheries in Scotland, worth £100 per annum – although he would still have to 'come and lick some great mans arse' to get it. Engineers were ripe for head-hunting. By 1775, the Russians had tried to recruit both Watt and Black. In 1776, the Batavian Society snapped up Jabez Carter Hornblower to get his up-to-the-minute steam-engine expertise. Robison ended up teaching maths to Imperial Cadets in St Petersburg. In 1781, when the partnership with Boulton had yet to deliver, the Russian ambassador teased Watt with an offer to make him supreme director of mines. He declined, too busy perfecting the machine, working around Pickard, and looking for that elusive rotative engine that would earn him more than any mining job. It was the wrong time for a Russian sell-out.

Another mixed opportunity – selling the engine and selling out – came from France. In 1786, Boulton and Watt travelled together to Paris to consider a proposal that they might build Watt's rotative engines in France under an 'exclusive patent'. The proposal might bring vast financial rewards. But would Watt sell to France a power that offered so much to Britain's prosperity? Had he taken a decision purely on business terms, he might well have gone ahead. Had he not been attracted by the proposition, he would surely not have travelled to Paris. But Watt did not become part of an eighteenth-century brain-drain, even by proxy. He developed scruples against a plan contrary to his domestic interests.

One other factor had immense consequences for the way in which the engine gradually became a familiar feature of the industrial landscape. Until 1796, before they established a foundry at their Soho works, Boulton and Watt did not, and indeed could not, make entire engines. Before then, they produced the crucial 'valves' regulating the movement of steam around the engine's complex mechanism, and they oversaw the final on-site construction. That was all. Working to designs supplied by Boulton & Watt and often acting under licence to them, other manufacturers did the rest. With Wilkinson's precision cylinders and William Jessop's pumps, Watt engines were hardly the offspring of one company, let alone one man. Designers, co-ordinators and networkers, Boulton and Watt were left to 'make it happen', piecing together disparate skills and far-flung components from around the country.

Clearly, what underlay the effective dissemination of the new technology was the permanence of the firm 'Boulton & Watt'. The partners worked hard over a quarter of a century to establish the firm, to streamline its organisation, and to optimise its profitability. Sometimes this will to profit had unexpected by-products. Working frantically in the mid-1770s, Watt was swamped with correspondence. Boulton impressed his factory visitors by showing how much human labour and skill he could automate. Watt considered how he might save clerical labour by copying letters 'automatically'. Two months of experiments in 1779 brought a reliable process, embodied in a robust and portable machine, patented the next year, and even demonstrated to the Royal Society. Letters were to be penned in a special ink concocted by Keir. Thin, damp, absorbent paper pressed onto the original gave a permanent copy, without the wasteful, time-consuming intervention of a fallible human scribe. 'James Watt & Co.', manufacturers of ink and presses, made the copying machine an office standard for the next century.

At Boulton & Watt, however, the processes of production still harnessed human physique and intellect. Robert Owen, New Lanark mill-owner and proto-socialist, would remind enlightened industrialists not to forget about people, those perfectible 'animate machines', too often neglected in favour of their 'inanimate' mechanical counterparts. Many of Watt's Lunatic friends were taken with Rousseau's writings on education. Like the Lunatics Keir, Darwin and Wedgwood, Boulton and Watt investigated systems of technical training (or was it indoctrination?) for their children and for their workers.

Boulton and Watt carefully nurtured James Jr and Matthew Jr, designing minds destined to 'captain industry' and initiating them into the Lunar Society. At the same time, they had a burning desire to dampen unruly human actions. Their experimental Albion Flour Mill went up in smoke in 1791 when workers panicked about losing livelihoods to steam. In the same year, Boulton and Watt armed trusted Soho employees in the expectation of attack. Abhorring democracy, Watt saw in the 'Rabble of this country' a 'mine of Gunpowder that will one day blow it up'. If Watt had to have a workforce, better

Illustration 11. James Watt's garret workshop at Heathfield. The workshop has been reconstructed at the Science Museum in London and serves as a shrine to the inventor. (Source: Science Museum, London.)

make it educated, disciplined and docile. They wanted an efficient shop floor, reliable managers and a persuasive sales force with an international outlook. None of this happened by chance.

When the partners finally retired from Soho in 1800, Watt was increasingly to be found at Heathfield Hall, the 'country pile' he put up near Birmingham. Thanks to their very recent success in the law courts, both were extremely rich men. By 1800, Boulton and Watt knew that it was safe to pass the business down to their sons who, as partners in the firm since 1794, knew the business inside out. Watt fitted out an attic workshop with a lathe and a small furnace where he could tinker away. Every corner was crammed with tools and detritus dating back to the Glasgow days. Just like old times, there was a forge in the back yard. Retreating from the firm's day-to-day business also freed up time for the scientific investigations that he had never quite abandoned.

· CHAPTER IX ·

'MY DEAR *PHILOSOPHE*':
JAMES WATT, MAN OF SCIENCE

The international dissemination of the engine para-
lleled the rise of Watt's reputation as a man of science.
If Boulton and Watt were 'partners in industry', Watt
and Black remained 'partners in science' as they
chatted on things philosophical, from botany to
geology, well into the 1790s. From the provincial
springboard of the Lunar Society and with Banks
as benefactor, Watt bounded towards the arenas
of scientific London, from which the tentacles of
national science stretched outwards to empire.
Robison hazily recalled of Watt: 'everything became
science in his hands'. Hutton was, as usual, only half
joking when he compared Watt with the French
Enlightenment encyclopaedists, and dubbed him 'my
dear *philosophe*'.

Watt's investigations into the properties of steam
(especially latent heats and boiling points at different
pressures) continued on and off from the 1760s for
at least twenty years. These systematic experiments,
the alkali project, and the campaign for universal

standards of measurement both for science and – with 'horsepower' – for machines, were evidence of Watt's voyage from the 'mere mechanician', toying with the Newcomen model, to scientific engineer and man of science, pronouncing with authority on the workings of nature. One example, the 'water controversy', shows that winning scientific credit could be as bruising as winning a patent battle, and the parallels were not lost on contemporaries. Olinthus Gregory swiped at Watt as 'too much actuated by a spirit of monopoly for a genuine philosopher', and that must surely have stung. A foray into 'pneumatic medicine' stands as evidence that Watt's passage from mechanician to philosopher was a circuitous one, taking him far from steam, and leaving him to invest as enthusiastically in a failing project as in a successful one. Finally, we will consider whether altogether new sciences might arise from the bid to get to grips with steam.

Patenting Nature? Watt and the Water Controversy

When Boulton and Watt visited France to sell the engine, Watt was flattered to find himself among the scientific élite of Paris. He fraternised with the chemist Antoine Lavoisier (1743–94), only to see that connection emphatically severed when the French savant and tax-gatherer lost his head in the bloody aftermath of the revolution of 1789. The theoretical chemist

Claude Louis Berthollet (1748–1822) greeted Watt like a brother, and rushed to share the secrets of chlorine bleaching for textiles.

Watt's collaborators Black, Roebuck and Keir had all been concerned with the practical chemistry of bleaching, dyeing and printing fabrics. After Watt married Ann MacGregor, in 1773, he had a father-in-law in the business too. This branch of industrial chemistry was of extreme economic significance, given the rapid expansion – not to mention the mechanisation – of the cotton industry in Scotland. Watt was always alert to the money-generating potential of practical 'improvements' like Berthollet's, and he set to work to make the process profitable on a large scale for Britain, and for James Watt. He wanted to patent it with James MacGregor. But Berthollet, loving science more than fortune on this occasion, would insist on making his 'discoveries on it publick', much to Watt's irritation.

Not all of Watt's philosophical forays were so closely targeted on profit, however. Basking in that Parisian warmth, he could, just about, present another persona. All the time he was promoting the steam engine, Watt remained a practical chemist, supported by his Lunar Society friends. That meant dealing with the 'phlogiston theory'. 'Phlogiston' was an invisible and supposedly weightless substance at the centre of a complex, occasionally mysterious, and increasingly problematic theory of combustion (burning) and 'calcination' (the 'burning' of a metal).

J.J. Becher (1635–82) had proposed and G.E. Stahl (1660–1734) had refined the theory, according to which burning matter gave off phlogiston to leave ash or, in the case of a metal, a 'calx'. So in order to 'restore' the matter to its original state, or to regain the metal from its calx, the chemist had to find a way of recombining it with phlogiston.

Phlogiston theorists like Priestley worked hard to extend the range of explanations with virtuosic experiments. Some of the new 'airs', isolated in an elaborate research programme of pneumatic chemistry, had names to match. In 1771, Priestley distinguished a special kind of air that assisted burning and promoted life. Believing it to be atmospheric air without phlogiston, Priestley called it 'dephlogisticated air' (before Lavoisier, in his radical reform of chemical language, saddled it with the new-fangled name 'oxygen'). Thanks to the phlogiston theory, there was even phlogisticated dephlogisticated air.

'Airs' could be tightly bound to the theory in other ways. In 1766, the well-heeled and eccentric Henry Cavendish (1731–1810) isolated a new 'inflammable air' (Lavoisier's 'hydrogen'). Priestley was convinced that inflammable air, although not weightless, was probably an 'aerial form' of phlogiston itself. But these names and identifications were unstable, not least because, as chemists increasingly made measurement – and especially weighing – the key to an experimental procedure, they ran up against a problem. A burning substance gained weight. A metal calx, having lost its phlogiston, was heavier than the original metal. Did that mean that phlogiston had negative weight?

Nevertheless, the phlogiston theory provided a viable framework for describing how common 'air' was made up from distinct components. After twenty years spent studying steam, James Watt asked whether water itself might be converted into air or, more fruitfully, whether it might be compounded, or made up from, simpler substances like the new airs of pneumatic chemistry. Watt knew this was a 'bold' suggestion. According to Aristotle, earth, water, air, fire and 'ether' were the basic elements, or building blocks, of all material things on earth and in the heavens, impossible to resolve into simpler constituents. Although the precise definition of an element was hotly contested in the intervening millennia, water retained its elemental status.

It was bold to ask the question, but there was good reason. Recent experiments by Priestley left Watt thinking. The new science of electricity was opening up new scientific vistas. Igniting inflammable air and dephlogisticated air with a spark of electricity, as Priestley did in March 1783, produced a mist, which condensed to become a 'moisture or dew'. Its weight was equal to that of the ignited airs. There were similar experiments in Paris, undertaken on a massive scale, by trustworthy gentlemen, keen, like Lavoisier, to measure with the greatest accuracy. Uniting these two airs brought a liquid equal in weight to the original airs. No known test could distinguish that liquid from pure water.

Ideas started to fly around in conversation and in private letters, rather than in scientific papers with carefully established submission dates. In April 1783, Watt wrote a letter to accompany the reports of Priestley's experiments, and stated his claim that liquid water was a compound of the two 'different kinds of air'. Watt wanted his written explanation to be given to Banks, as President of the Royal Society, so that it might be 'read' at a meeting of the Society. But the exact nature of his claim complicated matters: water was a combination of 'dephlogisticated air and inflammable air, or phlogiston' deprived of part of their latent heat (using Black's term). There was a problem. It turned out that Priestley's 'inflammable air' was not at all the same as Cavendish's. Because of the way in which Priestley obtained it, it might be a mixture of several different kinds of air. (In modern terms, his experiment was not about pure hydrogen and pure oxygen at all.) Watt's conclusion was at best ambiguous. Understandably, he wanted Banks to hold fire until more experiments clarified the matter.

In 1784, Watt and Boulton both became Fellows of the Royal Society of Edinburgh, a national rival to the Royal Society of London. Although there was heckling from communities of low-status mathematical practitioners, and provincial groups more closely aligned to practical utility, it was the London society which claimed convincingly to represent

British imperial science. It was not until April 1784 that Banks formally revealed Watt's views on the 'constituent parts of water'. With published papers in the *Philosophical Transactions*, the Society's venerable record of its miscellaneous researches, Watt achieved the coveted prize of a fellowship in November 1785. Boulton made it in the same year, and they both took the opportunity for a 'turtle feast' in the Society's Dining Club.

That achievement reminds us to place Watt in the company of his peers. Look at the men, all Fellows themselves, who supported this erstwhile 'mechanician' in his election bid. There was the illustrious Smeaton; there were Lunar Society friends Priestley, Whitehurst and Wedgwood; Henry Cavendish vouched for Watt. That Watt had the backing of the Secretary of the Levant Company, the Hydrographer to the East India Company and the Governor of the London Assurance Company indicates that the Royal Society was a place in which science, commerce and empire met. It also indicates that Watt, a legend in his own lifetime as 'inventor of the steam engine', was no lone genius at odds with the world.

But it was Watt's skills as an inventor, as much as his skills as a philosopher, that elevated him. The true 'discoverer' of the composition of water, after all, remained a subject of debate. Although Watt's letter had been read, belatedly, in April 1784, in the previous January the Royal Society had already heard similar claims from Henry Cavendish based on a similar sparking experiment but with pure dephlogisticated air and pure inflammable air. In the years to come, there was bitter wrangling over who (Watt, Cavendish or even Lavoisier) had first demonstrated the composition of water from oxygen and hydrogen, and whether the protagonists had acted independently. Priority disputes in science could be gruesome. Famously, the dispute between Leibniz and Newton over the invention of the calculus raged long after the protagonists' deaths. They were no less fierce than Boulton and Watt's battles with 'pirates'. In the culture of engineering, injured parties turned to law and to influential friends and patrons, whereas in science there was no simple court of appeal to decide the case once and for all. In science, powerful friends, or even national societies, would arbitrate over the all-important apportionment of 'credit' – all-important, since priority, once sanctioned by the scientific community, delivered reputation and that was what Newton, and even Watt, desired.

But those friends and societies had their own agendas. Cavendish's camp insisted that *he* had deduced the composition of water independently. Decades after the events took place, Joseph Black's student Henry Brougham concluded that the Secretary of the Royal Society had deliberately doctored the records before putting them in print, interpolating material to make Cavendish's claim more convincing. In 1834, the Parisian chronicler of scientific lives, Arago, made Watt the original and independent discoverer. Allies published documents designed to settle the dispute in Watt's favour once and for all. They did not. Now it is next to impossible to adjudicate on a dispute which tells us more about the protagonists than the elusive 'facts of the matter'. One authoritative modern commentator gives Watt priority only for stating that water was not an element; on this reading, it was Lavoisier who clearly specified its precise components. Whatever the truth of the matter, science surely was a 'litigious lady'.

Laughter in a Box: James Watt, Thomas Beddoes and 'Factitious Airs'

Science spilled over into medicine too. One branch of pneumatic chemistry, known as 'pneumatic medicine', was all the rage in the 1780s and 90s, a hot topic among chemists and medics in the Lunar Society and Priestley's friends in the Manchester Literary and Philosophical Society. They worried about the effects of bad air, in the growing towns, on aristocratic patients – and on vulnerable workers made distressingly unproductive by disease-carrying emanations. Watt was endlessly agitated about his aching head and spasmodic indigestion, and about his friends' 'several complaints' (Hutton's nasty urinary retention, Black's feverish attacks). Friends in the Lunar Society were dying of consumption; Edgeworth's wife and daughter succumbed. Two of Watt's children were also consumptives. When his 'amiable and lovely' Jessie died in 1794, he bewailed the 'sorrow that has unfortunately fallen to my lot'. Frantic to save his boy Gregory, 'a youth with every grace and virtue crowned' (so versified Keir), Watt turned quack, channelling his volcanic energies into the latest medical fad.

Pneumatic medics, like Darwin's friend Dr Thomas Beddoes, claimed they could promote happiness and cure disease by getting patients to inhale specially formulated 'factitious airs', that is, mixtures of airs (or gases) unlike atmospheric air. Since Beddoes' 'cloven Jacobin foot' was all too visible to his Oxford employers made nervous after the French Revolution, this 'fat little democrat' (as Edgeworth called him) found himself with plenty of time to survey patients – and even to experiment on himself. Air high in oxygen was good for the blind, the deaf and asthmatics like Watt. Air low in oxygen but high in hydrogen was the best thing for consumptives, although Black's 'fixed air' was worth a go too. If that failed, there was always 'hydro-carbonate', an impure form of carbon monoxide. This substance – 'a powerful anti spasmodick', Watt said – brought a healthy glow to the cheeks and had a wonderfully calming effect, although it did make one a little dizzy. But how to generate these airs? For carbon dioxide, Beddoes gamely suggested, take one cow, tether it outside your window and encourage it to breathe on you.

Looking for something better, James Sadler worked on the mass production of these airs – but his machines were expensive and inefficient. Watt now joined forces with Beddoes, and subjected Gregory to a regime of care which saw him shunted from the clear airs of Cornwall (Watt was there overseeing engines) to Beddoes' gaseous consulting room. With the resources of Soho behind them, Watt and Beddoes wanted economical production of factitious airs on a grand scale with optimal application to human lungs. The guiding principle, as with the steam engine, was to minimise waste and maximise effect, here the giving of health. The two part-time medical chemists devised apparatus to industrialise the generation of these airs for hospitals and to pump them out on a modest scale for the sickly amateur at home.

A pamphlet of 1795 explained how the apparatus worked, and totted up the cost to about £14. Boulton & Watt made the apparatus, complete with hydraulic bellows, oiled silk breathing bags, and shaped wooden mouthpieces. Driven by humanity, a tearful Watt had no wish to turn a profit here, and for once actively encouraged imitators in 'mechanical medicine'. (Perhaps they might realise Beddoes' dream of a mechanical leech?) Those too nervous to try the pneumatic apparatus at home could visit Beddoes' Pneumatic Institute, opened in Clifton, near Bristol, in 1799 with the hearty support of Wedgwood, Boulton, Darwin, Edgeworth, Keir and assorted family members. Watt was probably there to oversee the erection of the apparatus and, in another echo of the steam-engine business, he provided Beddoes with

William Clayfield, a skilled assistant, Soho-trained, to keep it up and running. Watt's favoured solution to the problem of gaseous delivery was a specially designed breathing chamber in which patients could be confined. Humphry Davy, Gregory's confidant and the Institute's superintendent, filled Watt's chamber with that 'heavenly air' nitrous oxide and ensconced himself to discover whether the new laughing gas gave him a hangover. (It did not.) Like the poet and addict Coleridge, Watt tried it, with disturbing effects. He had his own apparatus set up in the stables. Watt sent Black a kit so that he could administer the gas to a friend 'in a deplorable state of Hypochondriacism'. Much fun was had by almost all. Gregory died in 1804.

The Appliance of Science, or the Sciences of the Steam Engine

Watt was careful to insist that, in conceiving the separate condenser, he had not been directly guided by Joseph Black. Robison and Black begged to differ, but if, as seems likely, considerations of latent heat, and even of heat capacity, did not lead Watt directly to this elegant practical solution to the problem of wasted steam, then it is hard to claim that the steam engine was a simple application of the latest scientific theories from Glasgow to create a new technology. The situation is complex, however. Watt's friends called him a philosophical (or scientific) inventor.

After all, in the patent of 1769, Watt presented his improvements to the Newcomen engine not through a single new machine but in terms of guiding scientific principles.

If we understand science as so much more than a body of abstract theories, including instead all of the practical mess, material and apparatus associated with experiments, of the kind used by Priestley, Anderson and Black, then Watt's inventions were indeed scientifically framed. Watt was doing something rather unusual in Glasgow and Birmingham when, in trying to fix an engine, he trespassed on a world of systematic experimental investigation that was second nature to Black in his chemical investigations. In doggedly insisting that an existing practical technology, like an existing natural process, could be clarified using experiments, Watt represents a significant transition in the culture of mechanical engineering.

It is possible to turn this around further and say that the new Watt engine, and the many other engines working beside it, actually stimulated the creation of new sciences deliberately designed to understand, control, and then improve all kinds of heat-engines (like thermodynamics). To understand the Newcomen model, Watt himself had measured heat capacities and, effectively, come up with a figure for the latent heat of steam – although he needed Black to confirm that these phenomena of nature could be subsumed under an elegant scientific theory. To take another example, the practical idea of using steam 'expansively' (and economically) figured highly in France, although Watt was reluctant to adopt it. Sadi Carnot, one student of such engines, wrote in the abstract about how they might work in his *Reflections on the motive power of fire* (1824), and that book was of great consequence for William Thomson (Lord Kelvin), one of the founders of the new science of thermodynamics. A third example concerns Watt's parallel motion and his sun-and-planet motion. Such beautiful, but *ad hoc*, applications of geometry to practical mechanics were formalised in the mid-nineteenth century in a new engineering science of kinematics, which dealt with the geometry of complex machines in motion independently of any questions of force, resistance or work. In these cases, instead of a technology being a science applied to practice, we seem to have new sciences arising as technologies 'applied' to the science.

· CHAPTER X ·

THE PROGENY OF STEAM: PLANES, TRAINS AND AUTOMOBILES?

So far, the steam engine had its feet firmly on the ground. In the 1780s, engineers had transformed a pumping engine into a tremendously versatile source of power, but that source had always been fixed. Engines in the mines and factories were vast, and it was hard to imagine that any of their more diminutive progeny might be viable. But if they were, exciting futures beckoned in which steam might impel a boat, a carriage or even a flying machine. More often than not, those risky ventures were most clear to Boulton & Watt's competitors, or to younger men with less to lose. Yet these vivid futures, whether realistic or illusory, also raised the question of steam's longevity. If steam stood for progress, might progress leave steam behind?

Floating Steam

'Naval architects', or ship-builders, were quick to explore the potential of steam. With a high ratio of power to weight, Boulton & Watt's engines had the edge over predecessors and competitors. For the stationary engines immobile in mines and mills, this was of little consequence. For ships, it was vital. With a light but powerful engine in their grasp, engineers set to work to find ways of propelling boats through the water.

Engines *like* Boulton & Watt's looked set to engender radical new forms of marine propulsion. One idea floated by a talented rival has not quite been swamped by the partners' in-house historians. By 1787, the young William Symington (1764–1831) had patented a simpler engine than Watt's. It benefited from high steam pressures (like Hornblower's) and craftily avoided direct use of a truly 'separate' condenser. (Condensation took place in part of the working cylinder.) Hutton scorned it as an infringement, but Robison and Black, more realistically, grouched that Symington was sneakily sliding around Watt's patent. In any case, Watt and his staff were then busy 'realising' double-acting rotative engines for fee-paying factories, and they did not bother to oppose Symington.

Meanwhile, something mysterious was brewing and bubbling on the Firth of Forth. Watt's friends kept a weather eye, patent protectors turned would-be pirates, describing what they could see or only guess at. Symington's patron, the Edinburgh banker Patrick Miller (1730/1–1815), had been giving catamarans and trimarans paddle wheels for extra power when the wind dropped. He had already taken one of Symington's engines. Now Miller coupled the paddle wheel to a small engine in a double-hulled boat. Keen observers watched the world's first steam-powered vessel limping across Dalswinton Loch. If it made only five miles per hour, at least the boat remained un-incinerated. A fretful and forgetful Watt accused the engineer of sucking money from his patron. He bracketed him with his engine – 'of a piece, not very good' – and threatened, impotently, to 'speak to him in a legal way'. Hutton rejoiced that £4,000 had been spent on a boat which had 'a bad figure … cannot live in the sea … can carry nothing … can make no way'. But a year later, at Carron, of all places, a larger engine performed better for crowds of spectators, stretching the water-borne speed to half a dozen miles per hour.

When he came across problems, Miller turned to Watt for help. With crushing predictability, that collaboration foundered when it seemed that Miller and Symington were infringing Watt's patent. So much for Watt's gifts to humanity: only when the patent expired did Symington get back in the driving seat. In 1801, his 'steam-tug' *Charlotte Dundas* was churning the waters of the Forth and Clyde canal. Robert Fulton had witnessed Symington's original trials. He too waited, and between 1794 and 1805 dutifully placed orders with Boulton & Watt for engines to fit in the 'steamboats' he needed to take the United States by storm. It was a Boulton & Watt engine that powered Robert Fulton's 'steam-ship' (or

'S.S.') *Clermont* as it made a truly historic journey up the Hudson River in 1807. These early steamships relied on a great 'paddle wheel' for motion and manoeuvring. Eventually, the paddles that had replaced oars and sails would themselves be replaced by a curious device known as a 'spiral oar' (Watt's term). We might call it a 'screw propeller', or, having forgotten its origins, a propeller.

Given the lukewarm reception that Boulton & Watt gave to Symington and Miller, there is more than a little irony in the fact that the firm was taken to be so closely associated with the earliest experiments in marine steam. It was a short step to identify the name of Watt with the application of steam to 'naval architecture', especially in Scotland. Robert Napier, the originator of Clyde iron ship-building and marine engineering in the 1840s, had a prized portrait of Watt in his collection at West Shandon. There was no question for him who was the pioneer.

Driving Steam

As the new steamships and steam-powered canal boats made patently clear, not all steam engines need be stationary. When William Murdock, as Boulton and Watt's assistant, experimented with smaller and more efficient high-pressure steam engines, Watt discouraged him from obtaining a patent. Although Murdock put together a steam-powered model locomotive while he was working in Cornwall in

1784, his bosses persuaded him that he was wasting his time. With the benefit of hindsight, Watt missed a trick here. Perhaps it was because Watt's real concern, adopted from Boulton, was for brute power isolated in mills and mines, not for large sprawling networks of communication replacing the canals that he and so many others had surveyed. It was Edward Bull's collaborator Richard Trevithick who did most to mobilise Watt's rotative engines, making up his own railway locomotive carriage in 1804. Trevithick was another thorn in the tender side of Boulton & Watt, with his steadfast advocacy of using steam expansively, and at high pressures.

Managers of the new railway companies first turned, naturally, to their equine friends to tow railway carriages and freight trucks. Distinguished, self-publicising engineers like the insolvent Charles Blacker Vignoles and the big-thinking but diminutive Isambard Kingdom Brunel toyed with the idea of 'pneumatic railways', pushing carriages along with pressurised air confined to a track-side tube. As we know, these plans failed to oust the 'steam locomotive', a controversial technology (disparaged as a wasteful system of 'steam tugs') that nevertheless drove the British railway network and its maniacal expansion in the 1830s and 1840s. The steam railway, as a technological system integrating rolling stock, track and personnel, was again exported to Europe and far further afield. But that was not until years after Watt's death.

Many decades before the dense British railway network 'annihilated space and time', engineers considered how they might make a 'steam-carriage' for the common roads. Macadam and Telford were just two of the engineers deputed to manufacture a joined-up Britain, by consolidating her roads where canals were impossible and rivers un-navigable. Lunar Society members Edgeworth and Darwin built new carriages to suit these new roads. Darwin regularly used his to trundle from Birmingham to his Lichfield home. A popular physician, he trekked 5,000 miles a year 'imprizon'd in a post chaise', warring against pox and fever, and he wanted something better. Keenly aware of the stirrings in Soho, Darwin scribbled plans for a horseless carriage, one powered by steam. Independently of Watt, Edgeworth had been toying with the idea of moving land and water carriages by steam back in 1768. The ubiquitous Symington also put together model steam-carriages for the roads.

Where good roads existed, a steam-carriage would surely make an expensive, dedicated railroad redundant. Watt rarely missed such opportunities. He himself proposed a steam-carriage not for the 'rail-roads' that began to service the mines and factories, but for the common roads – including those which had been so improved during his century to promote economic regeneration in Scotland. As far back as 1758, when Robison and Watt first conspired to apply steam to practice, the goal was to give 'motion to wheel carriages'. Finally, in 1782, Watt patented just such a steam-carriage, as much to block other contenders as from any belief in a positive outcome. The real problem was to find a way in which a cumbersome carriage, weighed down with its own steam engine, could cope with the steep gradients of the British countryside. Watt's best and half-hearted attempt was to introduce special gears to regulate the delivery of power. Meanwhile, their faithful Murdock had become obsessed with an idea which, by 1786, Watt was convinced needed a miracle from God to work.

This was just an early example in a long history of engineers' attempts to woo the British public away from its canals, and later railways, onto the roads. The polite public squirmed at the prospect of destructive and noisy steam-carriages. How could a technology associated with heavy industry re-locate to the peaceful country road or crowded city street? But in the 1830s, Goldsworthy Gurney did manage to introduce a regular passenger route by steam-carriage from Gloucester to Cheltenham, which proved popular among his well-heeled public as an amusing novelty. Despite Gurney's best efforts, it never really took off. Vested interests squashed the idea, and there were intractable technical problems. Steam railways had been made to work only by radically changing environment and landscape, cutting up the countryside in a completely new way. Once the railways had taken off, steam-carriages trundling along rough and ready surfaces could not compete on grounds of price or speed. Constraints like these meant that some imagined steamy futures really were chimerical.

Superseding Steam

The very success of the steam engine made it fair game for all-comers. Steam challenged the familiar, and especially those classic 'prime-movers' dependent upon water, wind and animal power. But there were other contenders in the market for power. Paralleling Watt's 'improvements' to the Newcomen fire engine, European engineers hell-bent on novelty experimented with exploding gunpowder, pressurised carbonic acid gas (carbon dioxide) and outlandish 'internal combustions'. Seduced by 'something-for-nothing' schemes, the technological showman – or was it charlatan? – John Ericsson reasoned that if water and wind could not provide, better to harness the infinite power of the sun's rays. From time immemorial, vain and sometimes inane projectors (like Erasmus Darwin) had toyed and tinkered with minute mechanism, groping for the elusive secret of that perpetual motion machine that would bring about the ultimate revolution in technology. Steam really would be blown away.

One of the hottest contenders to supersede steam was the caloric engine. Many believed that heat was a weightless substance (caloric) and that the power of a 'fire engine' came from the application of this substance, working through and carried by a medium like steam. There was always the chance that caloric could be used and re-used, again and again, if only the right arrangement of mechanism could be conjured up. Often these engines relied upon the expansion and contraction of air rather than steam. Even the mainstream men of science who, in the mid-nineteenth century, decided that heat was motion and consolidated the splendid new science of energy – a science derived in large part from a desire to understand heat engines like the steam engine – were convinced that the caloric engine would ultimately 'supersede steam'. Irritated and inspired in equal measure by consummate publicists like Ericsson, engineers anticipated a new age in which a new prime-mover would make an old one redundant, pinching its markets in mines, factories and railways, displacing the steamship to 'cross the Atlantic by caloric', dominating the roads with caloric carriages, conquering the air with new-fangled flying machines, and offering the prospect of unlimited power. It hasn't happened – although the engineers are still working on it.

· CHAPTER XI ·

MONUMENTS AND MYTHS: RE-IMAGINING WATT

In 1781, Watt was struggling in his Soho workshop, frantic to universalise steam power, ranting at his rivals, worrying about empty order books and edging towards despair, despite Boulton's urgings that he pull himself together and succeed. After complaining for years of ill health, Watt was in his eighty-fourth year when he died on 19 August 1819 at Heathfield Hall, soon to be buried beside Boulton. By 1881, this same James Watt had been re-invented as a national paragon, celebrated as the genius of a revolution in power that promoted revolutions in industry and society, and distilled into the international language of science. We need to understand how this transformation from 'warts and all' mundanities to saintly transcendence happened.

The image-makers and literary executors set to work quickly, 'puffing Jamie' and puncturing his rivals long before he was cold in his grave. They have not ceased in their labours. Re-shaping the man, cleansing him of his sins, romanticising his personal history, canonising him, they have made a composite portrait: James Watt the solitary and unschooled craftsman, the inspired inventor, and the humble mechanician making it as gentleman philosopher. Here was an icon for his age. This secular saint has been a local hero, tied to Greenock, Glasgow, Scotland or Great Britain. He has his own icons in paintings, medallions and statues. Watt, set in stone, is a national asset. Adopted into science, he has become a universal possession.

Puffing Jamie

One well-tried and reliable strategy to secure a great posthumous reputation is to get admirers and acolytes to flood the market with sanitised histories and authorised, or doctored, biographies – or even to write them yourself. In the 1790s, Watt's friend Robison prepared perhaps the first historical note on the steam engine in his articles 'Steam' and 'Steam Engine' for the *Encyclopaedia Britannica*, then celebrated for its state-of-the-art knowledge. Watt found the articles wanting, especially where Robison talked about competitors or overplayed projects which did not bear fruit. When friends allegedly demanded that Watt revise them, he did so, claiming that it was a difficult task, since by 1814 he had not thought about steam for years, and had made the revisions only 'to my own great annoyance'. Still, he published a special limited edition for his influential friends in 1819.

It was an opportunity to make his own history: a

personal account of the invention of the separate condenser; a clarification of his role in establishing 'horsepower' as the standard of engine comparison; a final correction of Robison and Black's claims that, heaven forfend, he was the chemist's pupil and had followed his instructions to 'improve' the engine; and a surgical pruning of references to the ill-fated and best-forgotten 'steam wheel'. It was also a chance for a final stab at those Trumpeters and Imps of Satan, the Hornblowers.

Back in 1801, Watt had already modified Robison's surprisingly even-handed account of Jonathan Hornblower Jr and his rival engine. Specifically, Watt claimed that he (Watt) had used expansive action in 1775, a year before Hornblower's first experiments. (The true date of Watt's investigations was probably 1777.) He also said that, working before 1782, Hornblower had profited from (for which read 'borrowed ideas from') his 'intelligent employers' Boulton & Watt. Hornblower denied it. In the new revisions of the 1810s, Watt simply expunged Robison's account, his befuddled excuse being that Jonathan's machine had been found to be pirated (it had not).

Jonathan's mild-mannered attempts to hum his own praises were drowned out by the Watt propaganda machine. Jabez's accounts were suppressed by James Watt Jr. In a trend started, it seems, by Watt himself, J. and J. Hornblower have forever since been confused. Any 'company history' will highlight in-house achievements. Watt's partisan views were taken up in the earliest definitive histories of steam, written from 1827 by men like John Farey (1791–1851) – another friend and ex-employee. They have been broadly followed down to the present day. Competitors with few column inches, on the other hand, have been forgotten.

What could be done for the steam engine could be done for its putative inventor. In 1823, James Watt Jr dutifully celebrated his father's life, once again for the *Encyclopaedia Britannica*. Who knew him better, and who was less likely to cloud his image? Henry Brougham, by now a senior politician, penned a prose poem in which Watt stood for the white-hot fusion of natural philosophy and industry, a fitting emblem for his reforming Whig party.

French scientific administrators, too, were past masters when it came to hyperbolic *éloges* (or eulogies), embalming the dead bodies of national heroes in the fabulous liquor of scientific achievement, social improvement, and exemplary moral action. Watt's French connection as an associate of the Académie des Sciences earned him Arago's adulation in 1834, even as Watt's over-zealous propagandists played down French contributions to the engine's development. It is thanks to Arago's *éloge* that we hear, and hear, and hear again the quaint story of the teenage boy momentously inspired by that kettle as his foolish and short-sighted aunt chided him for time-wasting. If Newton had his falling apple, Watt must have his boiling kettle. As the philanthropic iron-master Andrew Carnegie (1835– 1919) thumped, 'let no rude iconoclast attempt to undermine' that compelling image, never mind whether it was true or false.

When physical infirmity denied James Watt Jr breath sufficient to puff up his father's achievements further, he called upon the Oxford-educated lawyer James Patrick Muirhead (1813–98) to write the definitive life. Such things were best kept in the family. Muirhead was a cousin of Watt (through the engineer's mother and his first wife), and for good measure he had married Boulton's grand-daughter. He 'Englished' Arago's memoir, tried to vindicate Watt as the discoverer of the composition of water, and spent a decade striving to give Watt's life and inventions adequate literary commemoration.

Muirhead started a trend. It has been hard, ever since, to separate Watt from the separate condenser. Samuel Smiles, prolific engineering biographer, continued another trend: airbrushing Watt's 'warts', extolling his real virtues and creating new ones, shrinking his business competitors, inflating his individualism, extolling his virtuous aversion to idleness, and making of him – or rather, the image of him – a thing to emulate in that greatest of Victorian goals: self-reliance.

Re-imagining Watt: Solitary or Social?

There are many answers to the question, 'Who was James Watt?' Every occupation he had has inspired its

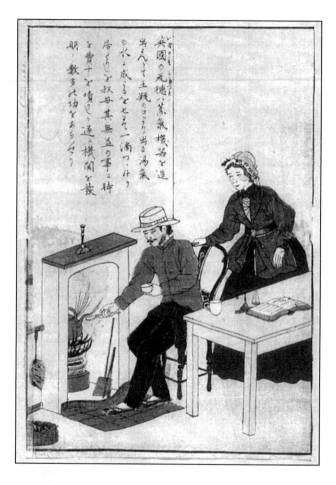

own image in the popular consciousness, nourished by Smiles. There were so many to choose from, and each had at least a grain of truth. But the silences, too, spoke volumes. Watt was at one moment the humble mathematical instrument-maker and solitary craftsman, Christ-like at his bench, poring over some minute construction, unbuffetted by worldly things until inspired from above – almost without conscious thought – to re-invent the steam engine for mankind's benefit. Less inspirational were the images of Watt acting, at other moments, as ruthless workshop manager surrounded by skilled assistants, or as cut-price surveyor vexed by the 'uncertainty' that was the steam engine.

Watt has suffered a fate similar to that of Thomas Alva Edison, whose name perhaps more than any other has come to symbolise inventive 'genius', associated as it is with the phonograph, electric light bulb and even moving pictures. Yet neither worked in a vacuum. Where Edison, in his 'invention factory', was surrounded by support staff, Watt, too, had always relied on others. Family, connections, partners, friends, craftsmen, factory workers, agents, patrons, politicians, propagandists, lawyers and philosophers crowded Watt. This was the society of steam that made his achievements possible. Hard as they tried to embellish the image, Watt was no more a lone genius than Edison.

It was more comforting to think of him emerging from his College cocoon, metamorphosed from 'mere mechanician' to philosopher by chance through the catalytic influence of the Newcomen model. There was little scope here for the entrepreneur and man of business, alternating anguish with tenacity in Boulton's expansive company. This engineer was better remembered as a part-time natural philosopher, nurtured by the Lunar Society but rising through graft – or was it inherent genius? – to the giddy heights of élite science in Edinburgh, London and Paris. It was a common trajectory of social and intellectual upward mobility, from mathematical practitioner to 'natural philosopher', from local hero to international asset.

Put in His Place: the Topographical Watt

Image-makers have worked hard to 'place' Watt. In one sense, his influence is dispersed and dissipated. Fragments of surviving engines, or their housings, or their products, become objects of pilgrimage in every industrial city, marking the transit of genius. But in another sense, he can be found and re-found in towns, cities and nations.

James Watt is easily placed as a 'son of Greenock'. Watt did not forget his birthplace. He returned there as an engineer, bringing better water and a more productive harbour. In 1816, just a few years before he died, he gave the people of Greenock a library. Here, in practice, he fulfilled his promise of workers' education while signalling his own transitions from parsimony to philanthropy and from work to word. Bathing in the glory of its local hero, Greenock turned his birthplace into the 'James Watt Tavern' and, in 1854, founded its very own 'Watt Club'.

It was easier, though, to identify Watt with Glasgow, quietly ignoring the defection to Birmingham that made it all possible. It was in Glasgow College, after all, that Watt had tinkered with the notorious Newcomen model. It was there that he had worked as 'Instrument Maker to the University'. College folk, like Dick, Muirhead, Anderson, Robison and Black, had patronised, befriended, shielded, stimulated and mentored him. Surely, then, it was Glasgow that had succoured the nascent 'genius' of James Watt – or rather, for the many that still believed Black's stories about skirmishes with city guilds, the College itself, as a progressive subset of the city? Certainly, the College gave this most notable autodidact one of its highest honours in a doctorate of letters (LLD) in 1806.

When the British Association for the Advancement of Science visited the second city of the empire in 1840, James Thomson, brother of William Thomson (Lord Kelvin), staged an exhibition of mechanical models, local manufactures, geological specimens and works of art. All of this served to celebrate Glasgow's industrial prowess, economic progress and growth from a bustling market town into a major centre of heavy industry – of iron, railways and steam engines. Glasgow would soon be the world leader in marine engineering, and the Clyde the 'shipyard of Empire'. Star billing in Thomson's exhibition went to the university's little Newcomen model – now, and perhaps forever, famous as the focus of Watt's dogged attentions some 80 years earlier – and to the engine of Henry Bell's *Comet*, itself a symbol of the origin of steam navigation. The meeting's managers wanted 'the names of Watt and Glasgow' united in 'imperishable records'. In the railway age, they spoke of steam's achievement: abridging time and space, and doubling, no less, the value of human life. There, indeed, was a prodigious saving of waste. For the Victorians, Watt came to exemplify the belief that, in Glasgow, science promoted invention and invention promoted wealth.

Of course, these ceremonials designed to identify Watt with place, or to make a kind of 'topographical Watt', extended to Scotland itself. He has many times been entered in the pantheon of 'famous Scots', most famously by his compatriot and biographer Carnegie. No matter that the English, not always well informed, have been sniping since the 1820s about the 'affected adulation of his countrymen who now plume themselves with his merits, though at first they refused him even a locality for a humble shop'. Now,

Watt has become a very British brand, transcending Greenock, Glasgow College and city, and even Scotland. Banks tried to have Watt made a baronet, and although it was a step too far, British crown and government found other ways of pouncing on industrial success, honouring it – and therefore appropriating it. Almost as soon as he passed away, there was a movement to make a monument for him – and of him.

Set in Stone: the Monumental Watt

On 18 June 1824, men of science, politicians, educators and industrialists galore gathered to plan a tribute to James Watt as the personification of (what we now call) the Industrial Revolution. These manufacturers of memory wanted something spectacular, symbolic and sanitising. Sir Humphry Davy, once Beddoes' chuckling pneumatic assistant, had married well and now moved freely among the élite of science and society. For him, Watt was a man who, 'more than any ... of this age, exemplified the practical utility of knowledge ... and [thus] both multiplied and diffused the conveniences and enjoyments of human life'. Sir Robert Peel, Tory politician and staunch supporter of science applied to practice, claimed Watt as the basis of his own prosperity, of the systematic organisation of the nation's cotton industry, and of the nation's wealth. Meanwhile, the Earl of Liverpool asked George IV if the Treasury might release funds for the

Illustration 13. Watt set in stone. Francis Chantrey's marble bust, commissioned by Watt himself in 1814, and copied for his son in 1816. James Watt Jr presented a further copy to the Royal Society of London as part of a campaign to fix and propagate his father's image. Chantrey's larger-than-life statues of Watt also appeared in Greenock, Glasgow, Manchester and, amid much controversy, London's Westminster Abbey. (Original owned by Lord Gibson Watt; image reproduced in Alex Potts, *Sir Francis Chantrey 1781–1841: sculptor of the great* (National Portrait Gallery, 1980), p. 23.)

monument. The name of Watt would thus be linked to the state and, moreover, to the unpopular King. Subscribers flocked to mimic royalty and all shades of scientific and political complexion.

If the time had come to set Watt in stone, there remained the question of who would best be entrusted the task. One man had a head start. In 1814, Watt had paid Francis Chantrey (1781–1841) 120 guineas for a bust – of himself. Rennie quickly followed suit. Watt had plaster replicas of both in his garret workshop, and planned to make a smaller copy for the mantelpiece with an apparatus of which he was boyishly proud, invented for that purpose, and which he discussed with Chantrey. Another Smilesian self-made man, Chantrey had a penchant for gadgets enhancing the sculptor's precision, styled himself a labourer in art, and managed a studio likened to an 'art manufactory' by virtue of its streamlined production, especially of high-quality replicas. James Watt Jr ordered one of these in 1816, sent another to the Royal Society of London and, all in all, spent £4,000 propagating his father's form.

No lightweight, by 1824 this 'Reynolds of portrait sculpture' had captured kings (Georges III and IV), authors (Wordsworth and Scott) and politicians (Canning and Wellington), not to mention national heroes (Nelson) and men of science (Joseph Banks). When Chantrey finished a full-length statue of him for the Handsworth tomb that year, Watt was in splendid, if stony, company. Variants of this went to Glasgow and to Greenock, paid for by public subscription.

It was not until 1834 that Chantrey's larger-than-life statue, priced at a staggering £6,234. 6s., finally entered London's Westminster Abbey. Henry Brougham, now the champion of the political left, adorned it with an epitaph reminding onlookers that it was 'not to perpetuate a name which must endure while the peaceful arts flourish, but to show that mankind have learned to honour those who best deserve their gratitude'. The engineer was not impious (Boulton had reminded him to pray for success), but the Lunar Society was hardly a centre of orthodoxy, and friends like Hutton were ditching religion revealed in scripture. Nevertheless, the memorialising process placed Watt at the centre of the established Church of England. If Watt did not have God on his side, he did at least have institutionalised religion.

Placed in George Square, Glasgow's vast statue could offend no one, but in London's Abbey, eyebrows were raised. The place was filling up with memorial statues jostling for prominence. This Watt was massive, unmissable and controversial. It has had an erratic trajectory, ousted from Westminster, buried in the crypt of St Paul's in 1966, and now in Scotland. Perhaps this is a manifestation of British ambivalence about the practical arts in which they consider themselves to have excelled. After all, it might not be so good to have machines for *everything*. Our friend

Hutton renamed Watt's engine the 'muscular motion whereby all the several parts shall be performed of erection, intrusion, reciprocation and injection'! Presumably Watt could see the joke, since he later told a bemused engineer: 'Your cock is too small (pardon the hint).'

Practical utility, replacing human labour and skill with machine, might be the foundation of national economic prestige, but 'steam intellect' clouded the imagination and strangled the spirit. One commentator saw the looming statue of Watt as a 'colossal champion of a new plebeian art', vying for space with monarchs, aristocrats and warriors, representing 'revolutions in the whole framework of modern society' equal to any commemorated in the Abbey. Chantrey's monument thus symbolised revolutions in technology and society when political revolutions were sweeping across Europe.

It is the multiplication of such images, in words, oil or stone, and the public preservation of Watt's personal and industrial effects, that helps to keep him as a household name. Later in the century, another statue appeared, altogether serious in demeanour. It stood in the Oxford University Museum. The eminent Victorian John Ruskin added Watt to his pantheon of scientific saints, beside such icons as Hippocrates, Galileo, Leibniz and Isaac Newton. Newton had his inspirational apple, and Watt – although spared his kettle – carried dividers and a tiny model engine, thus identifying man and machine. As further evidence of this sanctification, the garret workshop of Heathfield has itself become a shrine since it was moved in 1924, with all 6,000 of its sacred relics carefully preserved, to London's Museum of Science and Industry.

In no way has Watt's immortalisation been more clearly achieved than in his adoption as the scientific unit of power. A vast consensus thus inducts a man into the common language. The British Association for the Advancement of Science had been created to lobby for the interests of collective scientific endeavour at home, but ultimately it spoke to an international audience. Each year, the 'British Ass' met in a different town for a pageant of science. In 1882, 63 years after Watt's death, the assembled men of science assigned Watt's name to the unit of power (i.e., the rate of doing work). From the 1880s, international congresses have jostled and conspired to immortalise their chosen national heroes and heroines by stamping their names on universal physical units and on the fundamental constants of nature. We still talk of Volta and the (electrical) volt and Ampère and the amp. Not all of these names have stuck. It took the leaders of the electrical profession, meeting in Paris in 1889, to promote Watt as the international unit of 'output' (i.e., power); a Chicago Congress coaxed governments to enshrine standards in statutes. With machines now rated in kilowatts, Watt's own 'horsepower' dwindled. Now we find 'Watt', or his abbreviation 'W', on every light bulb.

Watt With or Without Warts?

Watt has surely become a national hero and an icon. He has been stripped down, clothed with all the virtues of his era, and re-dressed to fit the fashion of subsequent eras. We should never underestimate Watt's achievements, but celebrity has its price. We should not be shocked by Watt's humanity. Saying that he is human is not the same as saying that he was born to do what he did. Hindsight is a wonderful thing – but Watt was neither blindly inspired nor consciously prophetic. His motivations were complex, his life was extraordinarily varied, and the chances he took were so unpredictable that we cannot see him following a path fixed at birth.

Yet his environment was important. Saving waste, solving the Newcomen puzzle and conceiving the perfect steam engine made excellent sense when idleness was shunned, classes needed their entertainments, and Scotland was pushing for industrial expansion. When it was too late to make a difference, Watt admitted that a host of mechanics might have stumbled on ideas similar to his. One contemporary thought Watt a puppet of his times who had merely concentrated the scattered truths lying all about him with 'some little additional discovery' of his own. But where individuals were important, they need not be selfless. As the Lunar Society showed so wonderfully, self-interest and national benefit need not conflict. Personal loyalty, patronage and commercial ruthlessness made a heady and potent mix.

Watt was utterly conscious of the fact that perfecting the engine, transforming an 'arrangement in his mind' into a living mechanical 'giant with one idea', meant a sometimes agonising personal metamorphosis. Re-inventing steam meant re-inventing James Watt. It meant finding allies like Roebuck, Small, Boulton, Murdock, who could do what he could not do. It meant entering a competitive world of heavy engineering which was neither entirely alien nor remotely easy. If Watt struggled, it was in the company of all-too-talented competitors struggling – often against the might of Boulton & Watt. If, as Watt claimed, 'in this enterprizing age' one's 'thoughts seem to be stolen before one speaks them', technological conservatism made commercial sense. Starting too many hares meant losing the biggest prize.

A wartless Watt would have been no Watt at all. A very human James Watt was zealous but also jealous, careful to limit Black's role in the genesis of the separate condenser, spying on competitors, persecuting pirates, even starving the embryos of ideas that threatened, in their maturity, to usurp his steam monopoly. Be wary of accepting all the fine stories. Pity him for his wretched health. Remember Boulton as he chivvied him along, Watt lurching into another fit of depression about life, the steam engine, and everything. If Watt had an astonishingly fertile mechanical imagination, that does not mean that all other inventors were dullards or thieves. If Watt's engine changed and even benefited society, that does not mean that he failed to line his own pockets. If Watt showed a special kind of dogged determination, that does not mean that it was Watt against the world. To conceive the perfect engine, and to father so many of its imperfect progeny, Watt had a keen ability to fight his corner amid the endless bustle of techno-logical life. After all, as he said himself, 'without a hobby-horse, what is life?'

GLOSSARY

Atmospheric pressure: the pressure exerted by the earth's atmosphere, usually measured in pounds per square inch.

Crank: a mechanical arrangement whereby reciprocating (back and forth) motion can be transformed into rotary (circular) motion.

Cylinder: a hollow tube of circular cross-section and normally made of metal, in which the piston moves as the engine develops power.

Double-acting: of a reciprocating engine, one in which both 'back' and 'forth' (or 'up' and 'down') movements of the piston develop power.

Duty: a measure of engine performance, understood as the number of pounds in weight raised one foot by one bushel (later, one pound) of coal.

Governor: a device for regulating the working speed of an engine automatically (without human intervention), often using centrifugal force (hence 'centrifugal governor').

Horsepower: a measure of engine performance, in

terms of weight raised to a height in a given time. Watt preferred the term 'effect' to John Smeaton's 'mechanical power' or the term 'work', which became common parlance in engineering circles only in the mid-nineteenth century. From 1784, Watt made one 'horsepower' the capacity of an engine to raise a weight of 33,000 pounds by a height of one foot in a time of one minute.

Latent heat: coined by Joseph Black for the heat hidden, or 'latent', in a substance when it changes its state from solid to liquid, or from liquid to gas, without any change of temperature being registered on a thermometer.

Phlogiston: a principle of 'inflammability' in chemistry, according to the 'phlogiston theory' developed by J.J. Becher and G.E. Stahl from the late seventeenth century but still current, if controversial, in the mid-eighteenth century.

Piston: a circular disc, attached at its centre to a rod (the 'piston rod'); the piston moves up and down inside the cylinder.

Pressure: a measure of the action of a substance (e.g., steam) as it pushes against a surface: here, usually, weight (in pounds) per unit area (in square inches).

Reciprocating: of motion, back and forth in a straight line; of an engine, transmitting power by this kind of motion.

Rotary: of motion, in a circle.

Rotative: of an engine, developing rotary power

directly, rather than through an intermediary machine (e.g., a waterwheel).

Single-acting: of a reciprocating engine, one in which only one movement of the piston (either 'up' or 'down') develops power.

Stroke: of a piston, the distance moved inside the cylinder; of a machine, the repeating cycle of operations.

Vacuum: a space empty of matter, against which, for example, atmospheric pressure might act.

BIBLIOGRAPHY

This bibliography lists works which I have found particularly useful and which, in some cases, I have followed closely in explanations of technical detail. Two hundred years of authorship has generated a vast but exasperating literature on Watt, which must be approached with extreme caution. Among the very best guides have been Eric Robinson, Jennifer Tann, David Bryden, Hugh Torrens, Christine MacLeod and, most recently, David P. Miller. The works of Richard Hills have of course been absolutely indispensable, not least in suggesting this book's title.

History of technology:

Donald Cardwell, *The Fontana history of technology* (London: Fontana, 1994) (includes a brief authoritative account of Watt's activities).

Enlightenment context:

David Daiches, Peter Jones and Jean Jones (eds), *The Scottish Enlightenment 1730–1790: a hotbed of genius* (Edinburgh: Saltire Society, 1996).

Jean Jones, Hugh S. Torrens and Eric Robinson, 'The correspondence between James Hutton (1726–1797) and James Watt (1736–1819), with two letters from Hutton to George Clerk-Maxwell (1715–1784)', *Annals of Science*, vol. 51 (1994), pp. 637–53, and vol. 52 (1995), pp. 357–82.

Peter M. Jones, 'Living the Enlightenment and the French Revolution: James Watt, Matthew Boulton, and their sons', *Historical Journal*, vol. 42 (1999), pp. 157–82.

Watt as instrument-maker, merchant and civil engineer:

D.J. Bryden, 'James Watt, merchant: the Glasgow years, 1754–1774', in Denis Smith (ed.), *Perceptions of great engineers: fact and fantasy* (London: Science Museum, 1994), pp. 9–21.

Richard L. Hills, 'James Watt, mechanical engineer', *History of Technology*, vol. 18 (1996), pp. 59–79.

Richard L. Hills, 'James Watt's barometers', *Bulletin of the Scientific Instrument Society*, vol. 60 (1999), pp. 5–10.

Stephanie Pain (after Michael Wright), 'The flute-maker's fiddle', *New Scientist* (9 March 2002), p. 48.

Peter Swinbank, 'James Watt and his shop', *Glasgow University Gazette*, vol. 59 (1969), pp. 5–8.

The Newcomen engine:

Svante Lindqvist, *Technology on trial: the introduction of* steam power technology into Sweden, 1715–1736 (Stockholm: Almqvist, 1984).

L.T.C. Rolt and J.S. Allen, *The steam engine of Thomas Newcomen* (New York: Science History Publications, 1977).

Larry Stewart, *The rise of public science: rhetoric, technology, and natural philosophy in Newtonian Britain, 1660–1750* (Cambridge: Cambridge University Press, 1992).

The route to the separate condenser:

Donald Fleming, 'Latent heat and the invention of the Watt engine', *Isis*, vol. 43 (1952), pp. 3–5.

Richard L. Hills, 'How James Watt invented the separate condenser', *Bulletin of the Scientific Instrument Society*, vol. 57 (1998), pp. 26–9; vol. 58 (1998), pp. 6–10.

Richard L. Hills, 'The origins of James Watt's perfect engine', *Transactions of the Newcomen Society*, vol. 68 (1996–7), pp. 85–107.

Richard L. Hills, *Power from steam: a history of the stationary steam engine* (Cambridge: Cambridge University Press, 1989).

W.A. Smeaton, 'Some comments on James Watt's published account of his work on steam and steam engines', *Notes and Records of the Royal Society of London*, vol. 26 (1971), pp. 35–42.

The Lunar Society:

L. Gittins, 'The alkali experiments of James Watt and

James Keir, 1765–1780', *Transactions of the Newcomen Society*, vol. 68 (1996–7), pp. 217–29.

Barbara M.D. Smith and J.L. Moilliet, 'James Keir of the Lunar Society', *Notes and Records of the Royal Society of London*, vol. 22 (1967), pp. 144–54.

Robert E. Schofield, *The Lunar Society of Birmingham* (Oxford: Clarendon Press, 1963).

Robert E. Schofield, 'The Lunar Society of Birmingham: a Bicentenary Appraisal', *Notes and Records of the Royal Society of London*, vol. 21 (1966), pp. 144–61.

Horsepower and the indicator diagram:

R.L. Hills and A.J. Pacey, 'The measurement of power in early steam-driven textile mills', *Technology and Culture*, vol. 13 (1972), pp. 25–43.

Kinematics and the geometry of machines:

Eugene S. Ferguson, 'Kinematics of mechanisms from the time of Watt', *Bulletin of the United States National Museum*, vol. 228 (1962), pp. 185–230.

Disseminating steam:

Jennifer Tann, 'Marketing methods in the international steam engine market: the case of Boulton and Watt', *Journal of Economic History*, vol. 38 (1978), pp. 363–91.

K. van der Pols, 'Early steam pumping engines in the Netherlands', *Transactions of the Newcomen Society*, vol. 46 (1973–4), pp. 13–16.

Patents and pirates:

A.N. Davenport, *James Watt and the patent system* (London: British Library, 1989).

E. Robinson, 'Matthew Boulton and the art of parliamentary lobbying', *Historical Journal*, vol. 7 (1964), pp. 209–29.

E. Robinson, 'James Watt and the law of patents', *Technology and Culture*, vol. 13 (1972), pp. 115–39.

Jennifer Tann, 'Mr Hornblower and his crew: Watt steam engine pirates in the late 18th century', *Transactions of the Newcomen Society*, vol. 51 (1979–80), pp. 95–109.

Hugh Torrens, 'Jonathan Hornblower (1753–1815) and the steam engine: a historiographic analysis', in Denis Smith (ed.), *Perceptions of great engineers: fact and fantasy* (London: Science Museum, 1994), pp. 23–34.

Pneumatic medicine and the water controversy:

F.F. Cartwright, 'The association of Thomas Beddoes, MD with James Watt, FRS', *Notes and Records of the Royal Society of London*, vol. 22 (1967), pp. 131–43.

J.R. Partington, *A history of chemistry* (3 vols, London: Macmillan, 1962), vol. 3, pp. 344–62.

D.A. and R.G. Stansfield, 'Dr Thomas Beddoes and James Watt: preparatory work 1794–1796 for the Bristol Pneumatic Institute', *Medical History*, vol. 30 (1986), pp. 276–302.

Posthumous reception and cultural context:

Christine MacLeod, 'James Watt, heroic invention and the idea of the industrial revolution', in Maxine Berg and Kristine Bruland (eds), *Technological revolutions in Europe: historical perspectives* (Cheltenham: Edward Elgar, 1998), pp. 96–115.

David P. Miller, '"Puffing Jamie": the commercial and ideological importance of being a "philosopher" in the case of the reputation of James Watt (1736– 1819)', *History of Science*, vol. 38 (2000), pp. 1–24.

Alex Potts, *Sir Francis Chantrey 1781–1841: sculptor of the great* (London: National Portrait Gallery, 1980).

Paul Tunbridge, *Lord Kelvin: his influence on electrical measurements and units* (London: Peter Peregrinus, 1992).

Engines and energy:

D.S.L. Cardwell, *From Watt to Clausius: the rise of thermodynamics in the early industrial age* (London: Heinemann, 1971).

Robert Fox, 'Watt's expansive principle in the work of Sadi Carnot and Nicolas Clément', *Notes and Records of the Royal Society of London*, vol. 24 (1969), pp. 233–53.

Ben Marsden, 'Blowing hot and cold: reports and retorts on the status of the air-engine as success or failure, 1830–1855', *History of Science*, vol. 36 (1998), pp. 373–420.

Crosbie Smith, *The science of energy: a cultural history*

of energy physics in Victorian Britain (London: Athlone Press, 1998).

General and biographical sources:

H.W. Dickinson, *James Watt: craftsman and engineer* (Cambridge: Cambridge University Press, 1936; new edn, Newton Abbot: David and Charles, 1967).

R.V. Jones, 'The "plain story" of James Watt', *Notes and Records of the Royal Society of London*, vol. 24 (1969), pp. 194–220.

James Patrick Muirhead, *The origin and progress of the mechanical inventions of James Watt, illustrated by his correspondence with his friends and the specifications of his patents* (3 vols, London: John Murray, 1854).

Eric Robinson and Douglas McKie (eds), *Partners in science: letters of James Watt and Joseph Black* (London: Constable, 1970).

Eric Robinson and A.E. Musson (eds), *James Watt and the steam revolution: a documentary history* (London: Adams and Dart, 1969).

Samuel Smiles, *Lives of the engineers: The steam-engine: Boulton and Watt* (revised edn, London: John Murray, 1878).

George Williamson, *Memorials of the lineage, early life, education and development of the genius of James Watt* (Edinburgh and London: Constable, 1856).

Further reading:

Arne Hessenbruch, *Reader's guide to the history of science* (London and Chicago: Fitzroy Dearborn, 2000).